The Urban Book Series

The Urban Book Series is a resource for urban studies and geography research worldwide. It provides a unique and innovative resource for the latest developments in the field, nurturing a comprehensive and encompassing publication venue for urban studies, urban geography, planning and regional development.

The series publishes peer-reviewed volumes related to urbanization, sustainability, urban environments, sustainable urbanism, governance, globalization, urban and sustainable development, spatial and area studies, urban management, transport systems, urban infrastructure, urban dynamics, green cities and urban landscapes. It also invites research which documents urbanization processes and urban dynamics on a national, regional and local level, welcoming case studies, as well as comparative and applied research.

The series will appeal to urbanists, geographers, planners, engineers, architects, policy makers, and to all of those interested in a wide-ranging overview of contemporary urban studies and innovations in the field. It accepts monographs, edited volumes and textbooks.

Now Indexed by Scopus!

More information about this series at http://www.springer.com/series/14773

Dorina Pojani

Planning for Sustainable Urban Transport in Southeast Asia

Policy Transfer, Diffusion, and Mobility

Springer

Dorina Pojani
University of Queensland
Brisbane, QLD, Australia

ISSN 2365-757X ISSN 2365-7588 (electronic)
The Urban Book Series
ISBN 978-3-030-41974-5 ISBN 978-3-030-41975-2 (eBook)
https://doi.org/10.1007/978-3-030-41975-2

This Springer imprint is published by the registered company Springer Nature Switzerland AG
The registered company address is: Gewerbestrasse 11, 6330 Cham, Switzerland

To Ken, obviously

Acknowledgements

I am exceedingly grateful for the help, inspiration, and support received along the way. Many thanks are owed to: Theresa Audrey Esteban, Prakirati Satasut, Rifki Akbari, and Wen Hao Lee for their vital fieldwork and translation assistance; all the study participants in Manila, Bangkok, Jakarta, and Kuala Lumpur, who gave up considerable time to be interviewed and provided invaluable insights into their city's workings; my employer, The University of Queensland, Australia, for providing an Early Career Research Grant which allowed me to conduct fieldwork in Southeast Asia; IUAV University in Venice, Italy, and the Pontifical Catholic University in Santiago, Chile, for hosting me as I was writing this book during my sabbatical; Dominic Stead at Delft University of Technology, The Netherlands, for introducing me, during my postdoctoral residency, to the fascinating subject of policy transfer; Margaret Deignan, Sanjiev Kumar, and the rest of the editorial team at Springer for their patience and dedication in guiding this book from inception to publication; David Ashmore for providing feedback on an earlier draft of this book; the anonymous peer reviewers for their trust and encouragement; and my family in Albania and Australia, for everything.

Contents

About the Author

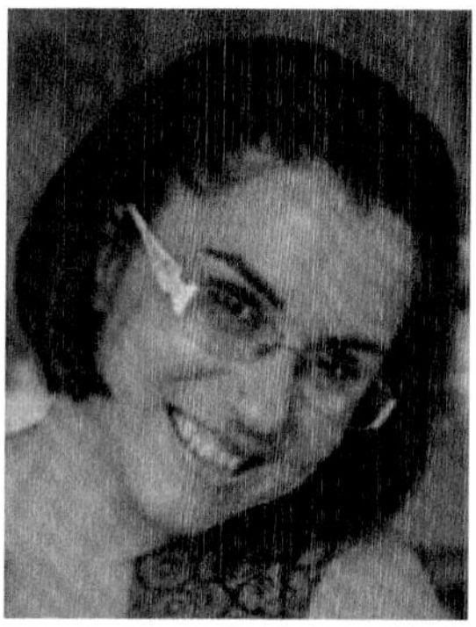

Dr. Dorina Pojani is Senior Lecturer in urban planning at the University of Queensland, Australia. Her research is international and comparative and encompasses built environment topics: urban transport, urban design, and urban housing. Other books by Dr. Pojani include *The Urban Transport Crisis in Emerging Economies* (Springer, 2017) and *Parking: An International Perspective* (Elsevier, 2020).

Author photo by Jetmir Veipi

List of Figures

List of Tables

Chapter 1
Policy Transfer as an Emerging Field of Study

Abstract This chapter introduces the book, which examines issues related to policy transfer in urban transport planning in Southeast Asia. The metropolitan regions of four major capitals—Jakarta, Kuala Lumpur, Manila, and Bangkok—are considered. The book assesses the in-bound and out-bound transfer of sustainable transport planning policies, concepts, and tools. The investigation focuses on who transfers policy and why, what elements of policy are transferred, in what direction and to what degree, and what barriers does transfer face. The analysis draws on eighty in-depth interviews with transport planners and policy makers from various government levels, non-profit organizations, private consultancies, academic departments, and international advisory agencies located in the case study cities.

Keywords Policy transfer · Southeast Asia · Urban transport · Jakarta · Manila · Kuala Lumpur · Bangkok

By now, planners everywhere know—more or less—what the ingredients of a sustainable city are, in theory (Pojani and Stead 2015). Everyone with a planning education, as well as a portion of the general public, has heard of concepts such as the transit metropolis, Copenhagenizing, the 30-minute city, the walkable city, the happy city, the smart city, the green city, the zero-carbon city, and the like. The problem is that only bits of solutions are being implemented in the cities that most need them, the majority of which are located in the Global South.

This book examines issues related to policy transfer in urban transport planning in Southeast Asia. The metropolitan regions of four major capitals—Jakarta, Kuala Lumpur, Manila, and Bangkok—are considered. The book assesses the in-bound and out-bound transfer of sustainable transport planning policies, concepts, and tools. The investigation focuses on who transfers policy and why, what elements of policy are transferred, in what direction and to what degree, and what barriers does transfer face. Is it simply the case that poorer cities lack funds for sustainable transport interventions? Or do barriers depend on the individuals and institutions involved in policy transfer, and their motivations and interactions? May cities be looking for inspiration in the wrong places? Or, are they looking in the right places but drawing the wrong lessons? Are they perhaps 'skimming' during policy learning from elsewhere?

D. Pojani, *Planning for Sustainable Urban Transport in Southeast Asia*, The Urban Book Series, https://doi.org/10.1007/978-3-030-41975-2_1

In the case of ex-colonies, up until independence a one-way exchange has typically occurred between the metropole and the colony, the former having traditionally served as a role model for the latter (Ward 2003). Is this type of relationship still in existence? How about neo-colonial relationships and new forms of coercive policy transfer—such as those reinforced by large international organizations and development banks (Woods 2006)? Are these present in Southeast Asia, and if so, how might they limit or bias the policy transfer process? Globalization may have worked to reinforce or weaken longstanding legacies. At the same time, it may have helped forge new allegiances. Are Southeast Asian capitals serving as role models for each other and/or for other less developed places in Asia and further afield? The book seeks to explore these issues. It also discusses whether and how policy transfer processes in the transport planning arena can be improved.

While policy transfer activities have been present all along the history of civilization (Ward 2003), in recent decades they have soared to unprecedented levels. In the West, where most planning research is still concentrated, planners recognize that much policy development and change is now prompted by transfers from elsewhere rather than internal innovations or pressures. At this point, the core activities of major international organizations such as the World Bank, the United Nations Development Programme, and the International Monetary Fund involve transferring planning policies from country to country or from city to city (Woods 2006). Within academia, policy transfer is becoming established as a bona fide field of study; it may even be considered as a new strand in planning theory (see Evans 2009).

Notwithstanding major efforts and substantial amounts of money spent on international consultants and policy advisors, existing surveys of the literature suggest that policy transfer in the transport arena has had a less-than-ideal success rate; irrational or unpredicted outcomes abound (Marsden and Stead 2011; Stead 2012). Despite difficulties, it is also clear that planning ideas and practices continue to spread—and not just from the 'west to the rest' or from the Global North to the South but possibly in every direction (Healey and Upton 2010). Nonetheless, in the context of Southeast Asia—a region comprising millions of people—policy transfer efforts in the field of transport planning have not been systematically studied. Very little is known of international technical assistance outcomes here.

Given Asia's meteoric rise to prosperity, there is great value in understanding how and from where Southeast Asian capital cities draw lessons to manage existing transport systems, to promote innovative transport solutions, and to meet demand for transport access and affordability. As looking elsewhere for answers to local problems is de rigueur, it is important to study policy transfer processes in a variety of contexts, including from the vantage point of Southeast Asia. At present, transport planning consultants who may conduct advisory or collaborative projects in the region are often in the dark as to what to expect.

Among all planning policy transfer issues, the effectiveness of urban transport systems has become a central concern everywhere. Transport problems follow a perverse pattern: while education and healthcare tend to improve as cities grow wealthier, congestion, pollution, safety, and energy security worsen (Pojani and Stead 2015). In much of the Global South, a radical overhaul of urban transport policies and practices is necessary at this juncture, if countries are to achieve equity, accessibility, and sustainability (Pojani and Stead 2017).

1.1 Case Studies

The selected case studies for this book (Jakarta, Kuala Lumpur, Manila, and Bangkok) are located in the four so-called 'Tiger Cub' countries (HSBC 2012): Indonesia, Malaysia, the Philippines, and Thailand (Fig. 1.1). They are so named because they have followed a similar model of economic development to that pursued earlier by

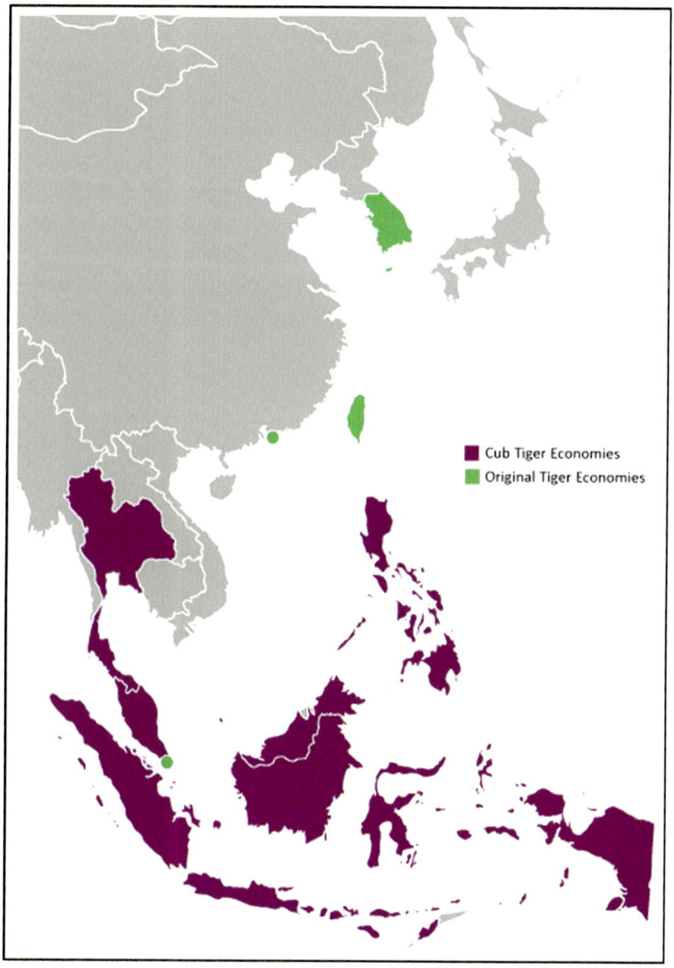

Fig. 1.1 Tiger Cub economies. Map by Zuanzuanfuwa (CC BY-SA 3.0), modified by author

Hong Kong, Singapore, South Korea, and Taiwan—the four original Asian Tigers.[1] All four Tiger Cub countries have large urban populations or are rapidly urbanizing; their economies are rapidly growing. Meanwhile, their transport-related problems—air, noise, and visual pollution, energy consumption, and traffic accidents—are severe. Economic growth has helped reduce poverty and increase the quality of life in these countries. However, at the same time, in all the capital city-regions economic growth and urbanization have placed major pressures on the transport systems and the populations that must navigate these systems.

The four capitals that serve as case studies have planning systems which are based on different social models, institutional setups, local cultures, and legal foundations (Table 1.1). Their levels of motorization differ too (Fig. 1.2). While Jakarta has more motorcycles than cars, Kuala Lumpur and Bangkok have some of the highest car ownership rates in Asia, rivalling Western Europe. Manila is considerably less motorized than its peers, but it is choked with traffic. These contextual differences offer an opportunity to examine issues of sustainable transport planning across a range of conditions. International comparisons are valuable in that they illuminate factors which may be overlooked in single case studies.

Differences notwithstanding, owing to their geographical proximity, the four case studies also share some historical patterns, cultural traits, and built forms. They are bustling, crowded, and congested metropoles, with temples, skyscrapers, factories, malls, shops, and homes placed alongside roads, highways, rail lines, rivers, and canals. Each capital is the centre of the respective country's socio-economic activity. Similarities tend to facilitate planning policy transfer between neighbouring countries (Stead 2012).

1.2 Data and Analysis

This book draws on evidence from in-depth interviews with transport planners and policy makers from various government levels, non-profit organizations, private consultancies, academic departments, and international advisory agencies located in Jakarta, Kuala Lumpur, Manila, and Bangkok. Eighty semi-structured interviews were conducted throughout 2016 (twenty in each metropolitan region). Participants were selected on the basis of their expertise on the topic, and involvement in various policy transfer activities. They were identified through snowball sampling (also known as chain-referral). This means that a few initial participants in each city were selected from within the researchers' professional network; upon completion of the interview, each person was asked to nominate other participants who met the eligibility criteria and could potentially contribute to the study.

The interviews lasted between a half hour and two hours. They were conducted in person, in the local language of each city, and were voice recorded. Upon transcription and translation into English, the analysis followed an iterative manual coding process

[1] More recently, Vietnam has been added to the Tiger Cub group.

Table 1.1 Overview of case study cities[a]

Jakarta Population: 10 million. Area: 255 square miles (611 square km). Transport system: Commuter rail, metro, light metro, Bus Rapid Transit (TransJakarta), bus, minivan (*angkot*), taxicab motorcycle taxi, auto rickshaw (*bajaj*), ride hailing (Go-Jek), and bikesharing (Gowes). Private motor vehicle trip share: 73% Jakarta is located on the northwest coast of Java on Jakarta Bay (Java Sea). Historically a swamp, the area floods easily during the monsoon season. The climate is tropical, with high temperatures and humidity throughout the year. Heat stress is exacerbated by high levels of air pollution. Jakarta was declared a 'special capital district' in 1966, thus gaining the status of a state or province. The metropolitan area (Jabodetabek) has a population in excess of 30 million, which is poised to grow despite government regulations prohibiting unemployed migrants from settling in Jabodetabek. Much of the population is young, resulting in a high natural growth rate. Dominating groups are of Javanese and Sumatran origin. Ethnic minorities include Arabs, Indians, and Chinese. Islam is the dominant religion. City planning has been influenced by both Dutch and British colonizers. The Dutch style of urbanism is evident in large, formal squares such as Medan Merdeka and Lapangan Banteng. Housing tends to be overcrowded. The most common type is the *kampong* (village) house, typically built of natural materials such as wood or bamboo. Government workers often live in 'colonial' houses (*rumah gedongan*), which are mostly single family. These are built in the 'oriental' style (*indische* style in Dutch), along wide, tree-lined streets with spacious gardens and lots. High-rise apartment buildings have been more recently introduced, and are concentrated in the financial district. While public transport is available, the city has prioritized the development of road and highway networks to accommodate private vehicles. As a result, traffic jams are commonplace (Waworoentoe 2019)
Kuala Lumpur Population: 7.9 million. Area: 94 square miles (243 square km). Transport system: Commuter rail, metro, light rail, monorail, buses, taxis, and bikesharing (oBike). Private motor vehicle trip share: 83% Kuala Lumpur is located in west-central Peninsular Malaysia, about 25 miles (40 km) east of the Indian Ocean, on the Strait of Malacca. The terrain is hilly and lies astride the confluence of the Klang and Gombak rivers. The climate is equatorial, with high temperatures and humidity throughout the year. Malays are the largest ethnic group but the city also encompasses large Chinese and Indian minorities. The majority of the population practices Islam; Buddhism has a large following too. Kuala Lumpur became the capital in 1957 (post-independence), and was designated a municipality in 1972. In 1974 the city and its surroundings became a federal territory, and the population surpassed one million by the early 1980s. The architecture comprises a mix of modernism and tradition; skyscrapers, shophouses, and stilt *kampong* houses coexist. Urban planning is influenced by Western, Middle Eastern, and East Asian styles. While the city centre along the Klang River is heavily congested, the metropolitan area is more rationally planned. Kuala Lumpur is the peninsula's transport hub, with rail lines and major roads radiating from it. The city has an extensive network of wide roads and highways, which, however, are congested. The light rail system was inaugurated in 1996. To cope with growing traffic congestion, at the turn of the twenty-first century a new administrative centre, Putrajaya, was built about 15 miles (25 km) south of the capital (Gorlinski 2019)

(continued)

Table 1.1 (continued)

Manila Population: 13 million. Area: 244 square miles (633 square km). Transport system: Light metro, commuter rail, buses, taxis, jeepneys, shared taxis (UV Express), ride hailing (Grab), bikesharing (ofo). Private motor vehicle trip share: 45% Located on the island of Luzon, Manila spreads over the low deltaic plain of the Pasig River, on the eastern shore of Manila Bay. The climate is tropical with only two seasons (wet and dry) but the city is protected from extreme weather conditions by nearby hills and mountains. Manila developed from an original fortress nucleus (*Intramuros*) and thirteen villages located outside its walls. The city and its satellites merged into Metro Manila (or National Capital Region) in 1975. The purpose was to provide integrated urban services (including transport) and to coordinate urban planning. Nonetheless, the region is now plagued with pollution, heavy traffic congestion, and overpopulation. It has a chronic housing shortage, although public housing has been built to shelter the poor and efforts have been made to resettle squatters into better quality accommodations. Building styles reflect colonial and neighbouring influences: American, Spanish, Chinese, and Malay, and range from traditional to modern. The population is predominantly Roman Catholic. Daniel H. Burnham, Chicago's prominent city planner, was active in Manila in the early twentieth century (Salita 2018)
Bangkok Population: 9.6 million. Area: 605 square miles (1,600 square km). Transport system: Metro, elevated train (Skytrain), Bus Rapid Transit, buses, taxis, minivans (*songthaews*), motorbike taxis (*motorsai taxi*), three-wheeler taxi (*tuk tuk*), river and canal boats. Private motor vehicle trip share: 57% Bangkok lies on the delta of the Chao Phraya River, about 40 km from the Gulf of Thailand. The climate is hot and humid throughout the year. The city did not become capital until the eighteenth century. In the contemporary era, it is characterized by sprawl and congestion. The centre, a formerly walled compound, is a major tourist destination. So are many Buddhist temples and monasteries (*wats*) located throughout the city. (Buddhism is the most common religion.) The inner city contains mostly institutional and commercial functions while the outer areas have a more residential and industrial character. Most floating markets of the past have now disappeared. Housing is generally overcrowded and consists of small, detached houses built on small lots with rudimentary sanitation facilities. Some high-rise apartment buildings exist around the centre, along with high-rise offices, hotels, and shopping centres. Many government agencies provide houses for their employees. Upscale housing for the wealthy, including the foreign community, is usually gated. While both the Thai government and the World Bank have invested in low-income housing construction, this has been insufficient to meet demand. Informal settlements are common on unused public land and along canals (*khlong*) (Sternstein 2019)

[a]Based on Encyclopedia Britannica

based on a combination of deduction and induction. By way of explanation, deduction is a top-down approach which aims at testing existing theory, whereas induction, a bottom-up approach, is concerned with the generation of new theory based on the data. In this case, the main themes were extracted from a pre-existing analytical framework (see below), whereas the sub-themes, which discuss substantive issues in Southeast Asian capitals, emerged directly from the data.

Before proceeding with the analysis, an overview of the theoretical approaches to studying policy transfer is provided below, together with a justification for the specific framework selected in this case (the 6D's).

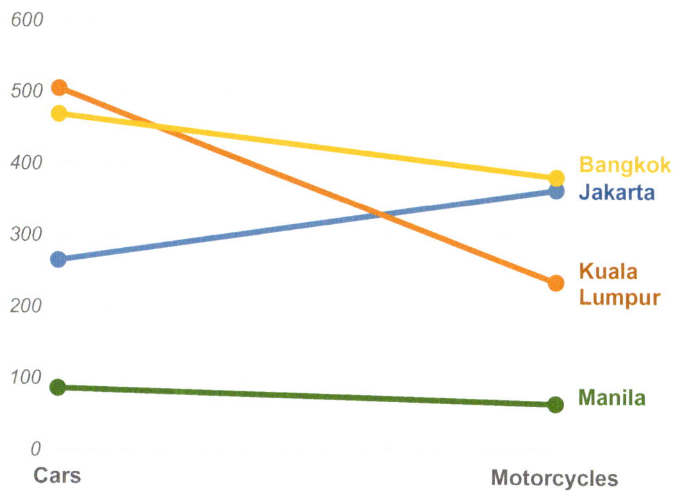

Fig. 1.2 Motorization in case study cities (vehicles per 1,000 inhabitants). *Data source* Leung et al. (2017) and country transport statistics

References

Evans Mark (2009) New directions in the study of policy transfer. Policy Stud 30(3):237–241

Gorlinski V (2019) Encyclopaedia Britannica. Kuala Lumpur. https://www.britannica.com/place/Kuala-Lumpur. Accessed 3 Aug 19

Healey P, Upton R (eds) (2010) Crossing borders: international exchange and planning practices. Routledge, London and New York

HSBC (2012) The World in 2050: from the top 30 to the top 100. Report. www.google.com.au/search?q=www.hsbc.com/~/media/HSBC-com/about.../the-world-in-2050.pdf. Accessed 7 Oct 2015

Leung A, Burke M, Yen B, Chiou YC (2017) Benchmarking urban transport oil vulnerability in 11 Asia-Pacific cities. J East Asia Soc Transp Stud 12:1005–1022

Marsden G, Stead D (2011) Policy transfer and learning in the field of transport: a review of concepts and evidence. Transp Policy 18:492–500

Pojani D, Stead D (2015) Sustainable urban transport in the developing world: beyond megacities. Sustainability 7:7784–7805

Pojani D, Stead D (2017) The urban transport crisis in emerging economies: an introduction. In: Pojani D, Stead D (eds) The urban transport crisis in emerging economies. Springer, New York, pp 1–10

Salita D (2018) Encyclopaedia Britannica. Manila. https://www.britannica.com/place/Manila. Accessed 3 Aug 19

Stead D (2012) Best practices and policy transfer in spatial planning. Plan Pract Res 27(1):103–116

Sternstein L (2019) Encyclopaedia Britannica. Bangkok. https://www.britannica.com/place/Bangkok. Accessed 3 Aug 19

Ward S (2003) Re-examining the international diffusion of planning. In: Freestone R (ed) Urban planning in a changing world: the twentieth century experience. E and FN Spon, London, pp 40–60

Waworoentoe WJ (2019) Encyclopaedia Britannica. Jakarta. https://www.britannica.com/place/Jakarta. Accessed 3 Aug 19

Woods N (2006) The globalizers: the IMF, the World Bank, and their borrowers. Cornell University Press, Ithaca, NY

Chapter 2
Theoretical Approaches to Studying Policy Transfer

Abstract This chapter discusses the various theoretical approaches to studying policy transfer. These include: policy transfer; policy mobility; policy convergence; policy diffusion; lesson-drawing; policy learning; bandwagoning; and knowledge transfer. The chapter also presents the six-faceted analytical framework employed in the study. The framework, adapted by the author, was originally proposed by David Dolowitz and David Marsh (2000).

Keywords Policy transfer · Policy mobility · Policy convergence · Policy diffusion · Lesson-drawing · Policy learning · Bandwagoning · Knowledge transfer

At this point, the policy transfer literature is sufficiently large to allow categorization. It can be organized into the following eight theoretical approaches (Table 2.1). These approaches overlap to some extent; however, there are important differences between them as well (Peck and Theodore 2010).

Policy transfer is concerned with how and why 'knowledge about policies, administrative arrangements, institutions and ideas in one political system (past or present) is used in the development of policies, administrative arrangements, institutions and ideas in another political system' (Dolowitz and Marsh 2000, p. 1). Early studies of policy transfer, which were mostly performed in liberal democratic contexts, viewed the process as rational and voluntary (Evans 2009a; Stone 2012). They also assumed a linear transfer between two places with a bilateral relationship. More recently, the literature has evolved to acknowledge that transfer can also be coercive or conditional to varying degrees (Stone 2012; Evans 2009b). In most cases, cities or countries have been found to borrow from others in a selective rather than indiscriminate manner, and transferred policies have been subject to interpretation and adaptation, resulting in hybrids that better fit local contexts (Stone 2012). Cases of poor, incomplete, partial, or failed transfer are now being documented, in addition to success stories (Thomas et al. 2018; Pojani and Stead 2018b). Finally, there is recognition that transfer can take place in multiple directions and iterations, thus eroding the distinction between policy 'lenders' and 'borrowers' (Stone 2012; Pojani and Stead 2014; Pojani and Stead 2015).

D. Pojani, *Planning for Sustainable Urban Transport in Southeast Asia*, The Urban Book Series, https://doi.org/10.1007/978-3-030-41975-2_2

Table 2.1 Different approaches to study policy transfer

Concept	Discipline	Authors	Interest	Focus
Policy transfer[a]	Political science	Dolowitz and Marsh (2000), Evans (2009a, b), Marsh and Sharman (2009), Stone (2012)	How is transfer of policy taking place, what is being transferred, and when it can lead to policy failure	Content, process, agency
Policy mobility	Geography	Cook and Ward (2011), Robinson (2011),Gonzalez (2011), McCann and Ward 2012	Linking global circuits of policy knowledge to local policy practice, politics, and actors	Process, agency, context, cities
Policy convergence	Political science	Bennett (1991), Howlett (2000)	The process of 'becoming' rather than 'being' more alike	Process, networking
Policy diffusion	Political science, planning	Berry and Berry (1990, 1999), Healey and Upton (2010)	How policies spread by osmosis or contagion	Adoption pattern
Lesson-drawing	Political science	Rose (1991, 2004), Majone (1991)	Under what circumstances and to what extent can a programme that is effective in one place transfer to another	Content, process
Policy learning	Political science	May (1992), Hall (1993), Dunlop (2009)	How do policy actors learn from abroad; who facilitates this	Process, epistemic communities
Bandwagoning	Political science	Ikenberry (1990)	Why do policies become fashionable and popular and attract increasing support	Process
Knowledge transfer	Organization studies	Argote et al. (2000)	How knowledge is transferred across organizational boundaries or within organizations; translation of ideas rather than copying	Process

[a]Concept and framework selected for this study. Table courtesy of Dominic Stead and Marcin Dabrowski (modified by author)

Policy convergence is defined as the tendency of planning systems, processes, instruments, and outcomes to grow more similar over time. The core characteristic of convergence is its dynamic rather than static nature. A mere comparison of cities or countries which are similar at a given point in time does not constitute policy convergence analysis. Converging implies 'moving from different positions toward some common point' in between (Bennett 1991, p. 219). In some cases, convergence may also imply unilateral movement of only one of two parties towards the other (Howlett 2000). Historically, theorists in the convergence camp have focused on explaining the notable convergence of industrial societies. One argument has been that a similar level of economic development produces comparable problems. In turn, comparable policies and political mechanisms are produced, leading countries with contrasting cultural and political traditions to becoming more alike (Bennett 1990). Convergence across places may occur intentionally, under conditions of shared information—for example, if one party has been selected as a benchmark of achievement for the other. Or, it can occur in parallel but independently or idiosyncratically—which makes it difficult to trace (Howlett 2000). Negative lessons may also be drawn from experience elsewhere, which then contribute to divergence rather than convergence (Stone 2012). Interest in policy converge re-emerged in the 1990s with the creation of the European Union, and the subsequent need for harmonization of policies in member countries.

Policy diffusion is sometimes categorized separately as a process of interdependent policy convergence—or policy change that occurs by osmosis (Stone 2012). Some scholars argue that policy transfer is a type of diffusion while others see diffusion as a type of policy transfer (Marsh and Sharman 2009). Policies diffuse as places strive to learn from, compete with, and conform to, one another. Geographical, ideological, and cultural proximity tends to assist diffusion (Berry and Berry 1990, 1999; Majone 1991). Socialization into shared norms by common training (e.g., in planning schools) or by interaction with professional organizations (International Society of City and Regional Planners, American Planning Association, Planning Institute of Australia, and the like) leads to policy diffusion as well (Berry and Berry 1999). Historically, diffusion has been explained according to a leader-laggard model, which assumes that certain places, typically more developed ones, are pioneers in the adoption of a policy, and other, less developed, places eventually emulate these leaders (Berry and Berry 1999). More recently, an understanding has emerged that 'there are all kinds of cross movements' of planning ideas (Healey 2010, p. 7). No one doing planning work can avoid being influenced, directly or indirectly (Healey 2010), although some contexts are more absorbent than others (Stone 2012). An analysis of diffusion processes is not simply about how planning ideas spread. It involves 'complex processes of translation, interpretation and adaptation' (Healey 2010, p. 5), and the arising tensions and contestations, which may themselves diffuse (Healey 2010).

Policy mobility. Rejecting the hierarchical model of policy diffusion and the Western hegemony, this school of thought maintains that policy ideas have the potential to 'circulate in all sorts of different directions across the globe, stretching across a wide variety of different contexts' (Robinson 2011, p. 21). In a post-colonial world, it may not be desirable or feasible to talk about policy directionality or mimicry—e.g.,

Global North to South (see McCann and Ward 2011). Given multiple and overlapping circuits of knowledge rather than a single planetary circuit, it may be impossible to trace the trajectory or evolution of a policy or technique as it moves nations, regions, or cities. This is evidenced by the fact that city strategies across a wide range of contexts, both wealthy and poor, are quite repetitive in terms of themes and proposed solutions (Robinson 2011). Policy mobility proponents posit that the complexity of policy mobility flows is 'unmappable'. As such, some researchers consider this approach to be 'anti-theoretical'—although others have embraced it.

Lesson-drawing. This approach assumes that the 'utilization of pre-existing models … is a common feature of policy innovation' (Majone 1991, p. 103). In the case of intermestic issues (such as environmental pollution) which combine both international and domestic influences, lesson-drawing is paramount (Majone 1991). However, the object of lesson-drawing is not to copy programs or to theorize on possible solutions to a problem but to learn under what circumstances and to what extent effective programmes are transferable (Rose 1991, 2004). Lesson-drawing does not take place for its own sake or 'to collect exotic information' but it is instrumental in improving conditions in society and promoting political satisfaction (Rose 2004). 'A lesson is more than an evaluation of a programme in its own context; it also implies a judgement about doing the same elsewhere' (Rose 1991, p. 7). As such, it is an action-oriented conclusion. The failures of other places offer lessons too, and may help avoid newcomer costs (Rose 2004). Thus, external models can exercise both push and pull forces on policy (Majone 1991). Much of lesson-drawing theory, like comparative politics, focuses on differences between policies that different governments adopt in response to a common problem (Rose 2004). Unlike convergence or diffusion, this approach does not assume that there are common responses to common problems (Rose 1991).

Policy learning. This portion of the literature is focused on the psychological and cognitive dimensions of policy transfer (Hall 1993). Unlike mimicking or copying, policy learning implies improved understanding via experience and trial-and-error—although it 'does not necessarily entail discovery of an "objective" truth about a policy instrument, problem, or goal' (May 1992, p. 351). Learning can be incremental or radical (Hall 1993), and ideas can spill over from a policy domain to another—not only from one place to another (May 1992). Epistemic communities and cognitive elites are central here as they assume control over knowledge production and, in so doing, they intentionally guide decision-makers' policy learning (Dunlop 2009). However, policy learning is not necessarily synonymous with policy transfer or adoption because, alike lesson-drawing, learning may also be negative (when decision-makers learn what not to transfer) or even wrong (when the original context is misunderstood). Policy failure may also trigger learning (Dunlop 2009; May 1992). 'Learning may be more or less formal, and individuals' –preferences and values mediate what it is learned and from whom' (Dunlop 2009, p. 291). A policy learning approach 'raises questions about what happens over time as individuals' understandings mature or circumstances change and also when there are personnel changes in key decision-making positions' (Dunlop 2009, p. 307).

Bandwagoning is the process of joining others in adopting or supporting policies, tools, and activities which are fashionable or likely to be successful. For example, bandwagoning made privatization part of a new international policy culture in the 1990 s while the post-war era saw the bandwagon success of Keynesianism (Ikenberry 1990). Sometimes bandwagoning involves alignment with a stronger adversarial power. The appeal, stickiness, and ultimate influence of bandwagons might have more to do with their emotional quality than with reason or reality (Cox and Béland 2012).

Knowledge transfer literature focuses on organizations rather than planning processes, political systems, or entire societies. Knowledge transfer is defined as the process through which one unit in an organization (e.g., a planning department in a city) is affected by the experience of another (e.g., a planning department in another city). Empirical evidence suggests that organizations which are able to transfer knowledge more effectively from one unit to another are more productive and resilient than others (Argote et al. 2000). Beyond the sharing and adaptation of information, successful knowledge transfer involves 'storing this information so that it can persist in the face of individual turnover' (Argote et al. 2000, p. 5).

Generally, the foregoing literature is much stronger on hypothesis formulation than on systematic empirical analysis. Analytical frameworks which can operationalize theory and guide data collection have been rarely laid out. An exception is the study by Dolowitz and Marsh (2000), which has proposed a multi-dimensional policy transfer framework. This framework has been employed in a range of disciplines, including transport, in both Global North and South city contexts (see Pojani and Stead 2014, 2015, 2018b; Si et al. 2020; see also Benson and Jordan 2011). It is adopted in the present study as well.

The framework is appropriate for answering the research questions set forth earlier: who transfers policy and why, what elements of policy are transferred and to what degree, from where to where are policies transferred, and which constrains does transfer face? It covers three areas of study: (1) *description* of how policy transfer is made, (2) *analysis* of why policy transfer occurs, and (3) *prescription* of how policy transfer should be made (Evans 2009a, b).

The six facets of the framework (6D's, for ease of recall) are briefly presented below and summarized in Table 2.2 (see also Pojani and Stead 2014; 2015; 2018b):

(1) *Drives for policy transfer*. These vary from voluntary, perfectly rational, to coercion by advocacy groups, political parties, funding bodies, or policy entrepreneurs. Voluntary transfer tends to occur in more developed contexts while coercive transfer is more common in the Global South. Mixed forms of transfer are also undertaken in order to secure grants, loans, or other investments. Politico-economic crisis or image concerns may also provide a drive for policy transfer.
(2) *Drivers of policy transfer*. These may include politicians, civil servants, advocacy groups, think-tanks, consultants, academics, international organizations, and supra-national institutions. Each driver plays different roles and brings a different set of attitudes, values, and resources to the process.

Table 2.2 Analytical framework (6D's) for the study of policy transfer[a]

Drives (Why are policies transferred?)	Drivers (Who engages in policy transfer?)	Description (What elements of policy are transferred?)	Direction (In what direction are policies transferred?)	Deterrents (What constraints does transfer face?)	Degree (To what degree are policies transferred?)
Deliberation	Roles	Policies	Intra-national	Cognitive	Soft
Coercion	Attitudes	Goals	Cross-national	Structural	Hard
Mixture	Values	Tools	Own past	Technical/technological	Failure
Osmosis	Resources	Projects			
		Programs			

[a]Based on Dolowitz and March (2000), Evans (2009b), Pojani and Stead (2015)

(3) *Description of policy transfer*. What is transferred or sought to be transferred may include policies, goals, tools, projects, or programmes.
(4) *Direction of policy transfer*. Policies can be transferred within one country or across borders. A city's or country's own past may also serve as a source of ideas. The focus of this book is mainly on cross-national transfer, which can present different challenges and produce more diverse outcomes than policy transfer within a single country.
(5) *Deterrents to policy transfer*. These may be cognitive or structural. Cognitive obstacles may be due to an insufficient search for new ideas, ideological clashes between policy lenders and borrowers, a lack of contextual understanding on part of policy lenders, excessive complexity of the policies to be transferred, physical distance, and language barriers. Structural obstacles denote failure to effectively mobilize local elites, limited financial and human resources among borrowers, and/or a lack of cohesive policy transfer networks. Public opinion may also be a deterrent to policy transfer. This is shaped by the media, the elites, and constituency groups.
(6) *Degree of policy transfer.* 'Soft' policy transfers involve changes in ideas, concepts, and attitudes, whereas 'hard' policy transfers signify changes in programmes and implementation. Policy borrowers may copy or emulate a lender's policies, or simply be inspired by those. Negative lessons can also be drawn during and after policy changes, which may be as valuable as positive lessons.

The empirical research for this book was designed in accordance with the 6D's framework. The findings are discussed below.[1] Note that much of the discussion is based on the views and opinions of the interviewees.[2] This 'soft', qualitative research

[1]All photos that accompany the text are by the author, unless otherwise noted.

[2]The analysis makes extensive use of direct quotes from the interviews. The reader is reminded that quotes are not original but have been translated from local languages (seldom were the interviews conducted in English). Virtually all quotes have been edited for clarity, readability, and anonymity, while retaining the intended meaning. For example, 'filler' or repeated words have been removed, and sentences within a paragraph have been reordered to facilitate the logical flow of the discourse. Long sentences, with multiple dependent clauses, which work may well in speech but not in writing,

approach is intentional. It resembles planning- and policy-making processes, much of which depend on perceptions, values, political power, and popular pressures rather than 'hard' science. As philosopher Thomas Kuhn (1996) postulated, reality cannot be described independently of the conceptual schemes through which we observe it. Accordingly, much professional planning work is relativistic in that it accepts that there is no universal 'right' or 'wrong' for all people and places (Scott 1998), and embodies ambiguity regarding ends and means (Brooks 2002). This study, too, takes a pluralistic stance by giving voice to a multitude of actors who shape, or at least influence, transportation planning in Southeast Asian capital cities.

References

Argote L, Ingram P, Levine J, Moreland R (2000) Knowledge transfer in organizations: learning from the experience of others. Organ Behav Hum Decis Process 82(1):1–8
Bennett C (1991) Review article: what is policy convergence and what causes it? Br J Polit Sci 21:215–233
Benson D, Jordan A (2011) What have we learned from policy transfer research? Dolowitz and March revisited. Polit Stud 9:366–378
Berry FS, Berry WD (1990) State lottery adoptions as policy innovations: an event history analysis. Am Polit Sci Rev 84:395–415
Berry F, Berry W (1999) Innovation and diffusion models in policy research. In: Sabatier P (ed) Theories of the policy process. Co: Westview Press, Boulder, pp 169–200
Brooks M (2002) Planning theory for practitioners. APA Press, Chicago
Cook I, Ward K (2011) Trans-urban networks of learning, mega-events and policy tourism: the case of Manchester's Commonwealth and Olympic Games projects". Urban Stud 48(12):2519–2535
Cox RH. and Béland D (2013) "Valence, policy ideas, and the rise of sustainability." Gov: Int J Policy Adm 26(2): 307–328
Dolowitz D, Marsh D (2000) Learning from abroad: the role of policy transfer in contemporary policy-making. Gov: Int J Policy Adm 13(1):5–24
Dunlop C (2009) Policy transfer as learning: capturing variation in what decision-makers learn from epistemic communities. Policy Stud 30(3):289–311
Evans M (2009a) New directions in the study of policy transfer. Policy Stud 30(3):237–241
Evans M (2009b) Policy transfer in critical perspective. Policy Stud 30(3):243–268
González S (2011) Bilbao and Barcelona 'in motion': how urban regeneration 'models' travel and mutate in the global flows of policy tourism". Urban Stud 48(7):1397–1418
Hall P (1993) Policy paradigms, social learning and the state: the case of economic policy-making in Britain. Comp Polit 25:275–296
Healey P 2010 The transnational flow of knowledge and expertise in the planning field. In: Healey P, Upton R (eds) Crossing borders: international exchange and planning practices, Ch. 1. Oxon, UK: Routledge, pp 1–26
Healey P, Upton R (eds) (2010) Crossing borders: international exchange and planning practices. Routledge, London and New York
Howlett M (2000) Beyond legalism? Policy ideas, implementation styles and emulation-based convergence in Canadian and US environmental policy. J Public Policy 20(3):305–329

---

have been simplified. In a number of cases, when responses within the same city were quite similar, composite quotes have been created. To protect the interviewees' privacy, their personal data (name and gender), as well as their specific job titles, have been withheld in all cases. Only the city in which the interview originated has been noted at the end of each quote.

Ikenberry J (1990) The international spread of privatisation policies: inducements, learning and policy bandwagoning. In: Suleiman E, Waterbury J (eds) The political economy of public sector reform and privatisation. Co: Westview Press, Boulder
Kuhn T (1996) The structure of scientific revolutions. University of Chicago Press, Chicago, Il
Majone G (1991) Cross-national sources of regulating policy-making in Europe and the United States. J Public Policy 11:79–106
Marsh D, Sharman J (2009) Policy diffusion and policy transfer. Policy Stud 30(3):269–288
May P (1992) Policy learning and failure. J Public Policy 12:331–354
McCann E, Ward K (eds) (2011) Mobile urbanism: cities and policymaking in the global age. University of Minnesota Press, Minneapolis, MI
McCann E, Ward K (2012) Assembling urbanism: following policies and 'studying through' the sites and situations of policy making. Environ Plan A 44:42–51
Peck J, Theodore N (2010) Mobilizing policy: models, methods, and mutations. Geoforum 41(2):169–174
Pojani D, Stead D (2014) Dutch planning policy: the resurgence of TOD. Land Use Policy 41:357–367
Pojani D, Stead D (2015) Going Dutch? The export of sustainable land-use and transport planning concepts from the Netherlands. Urban Stud 52(9):1558–1576
Pojani D, Stead D (2018) When West-East planning policy transfer fails to gain traction. J Environ Plan Manag 62(8):1402–1419
Robinson J (2011) The spaces of circulating knowledge. In: McCann E, Ward K (eds) Mobile urbanism: cities and policymaking in the global age. University of Minnesota Press, Minneapolis, pp 15–40
Rose R (1991) What is lesson-drawing? J Public Policy 11(1):3–30
Rose R (2004) Learning from comparative public policy: a practical guide. Routledge, New York and London
Si K, Wereta T, Pojani D (2020) South-south policy transfers? Transit Oriented Development from China to sub-Saharan Africa". In: Chen CL (ed) Handbook on transport and urban transformation in China. Edward Elgar, London
Scott J (1998) Seeing like a state: how certain schemes to improve the human condition have failed. Yale University Press, New Haven, Conn
Stone D (2012) Transfer and translation of policy. Policy Stud 33(6):483–499
Thomas R, Pojani D, Leffernik S, Bertolini L, Stead D, van Krabbe E (2018) Is Transit Oriented Development (TOD) easily transferable to international contexts? Reg Stud 52(9):1201–1213

Chapter 3
Transport Policy Transfer in 'Tiger Cub' City-Regions

Abstract This chapter discusses the findings of the study. It is structured in accordance with the six facets (the 6D's) of the analytical framework set forth in the previous chapter: (1) drives, (2) drivers, (3) description, (4) direction, (5) deterrents, and (6) degree of policy transfer. Much of the discussion is based on the views and opinions of the interviewees. The analysis makes extensive use of direct quotes.

Keywords Policy transfer motivations · Policy transfer actors · Policy transfer description · Policy transfer direction · Policy transfer barriers · Policy transfer outcomes

The following discussion is structured in accordance with the six facets (the 6D's) of the analytical framework set forth above: (1) **Drives**, (2) **Drivers**, (3) **Description**, (4) **Direction**, (5) **Deterrents**, and (6) **Degree** of policy transfer. The key research questions are reiterated at the start of each section.

3.1 Drives for Policy Transfer

Why are policies transferred? Are motivations voluntary or coercive or a mixture?

The 'sustainability' chimera

The ideology of globalization, promoted since the early 1990s, has led (or perhaps forced) individual nations to open up to the world, and integrate into international networks—not only of trade but also of knowledge (Evans 2009b). Consequently, 'sustainability'—variably defined in environmental, social, or economic terms—has become a mantra everywhere, at least on paper. Unavoidably, the concept has spread by osmosis to Southeast Asia as well. Moreover, domestic transport problems such

D. Pojani, *Planning for Sustainable Urban Transport in Southeast Asia*, The Urban Book Series, https://doi.org/10.1007/978-3-030-41975-2_3

Fig. 3.1 Typical levels of traffic congestion and overwhelming transport infrastructure in Manila

as extreme congestion, pollution, and accidents have led local populations to demand more sustainable solutions (Fig. 3.1).

The dramatic quote below recounts how stressful city life becomes when the transport system is poor. In some ways, the transport-related stress is more salient now that populations are aware of better urban transport models elsewhere, and constantly compare their home city to those:

> Sustainable transport is a necessity, and ordinary people also want it… People living in Bangkok are sick of the traffic congestion. … Right now, they are forced to scramble for their own solutions… driving 3–5 h per day… At times, people seem ready to lash out at each other on the street… When I was in Japan, I could ride my bike instead… I did not need a car. Their mass transit system did not make transport a problem in my life. I did not have to plan much if I needed to travel… In Bangkok, if I have two errands to run, I have to think really hard about how to run those as fast as possible. It requires a lot of planning, which does not necessarily solve the problem. You can only plan so much, but if you get on an expressway and get stuck in traffic for two hours, then all that planning goes out of the window. … If you are poor, you are left with no choice but to walk on roads that are invaded by three-wheelers. … We cannot even figure out how to get to public parks. We do not know where they are, only that they are far away. We live in a stressful environment. The only the option we have is to either stay home or go to shopping malls (Fig. 3.2). … Transport is not just transport. It is a matter of life… Our lives are not taken care of. … (Bangkok).

Fig. 3.2 Upscale but faceless shopping malls are among the few car-free spaces in Southeast Asian capitals. The examples above are from Bangkok (right) and Manila (left)

At the same time, interviewees note that extreme transport conditions are prompting some action:

> We've now reached the level of 'emergency', so we can no longer use the usual priority scale [when deciding which sustainable transport projects to implement]. All have to be done (Jakarta).
>
> Because of climate change, over the last ten years they really are pushing some [sustainable transport] proposals. I don't know if anything has been achieved but they are talking the right talk (Manila).

In addition to domestic pressures, most Southeast Asian countries subscribe to various international treaties (e.g., the Kyoto Protocol[1] or the Paris Agreement[2]), which require signatories to take concrete actions to reach certain sustainable transport targets. Even if progress is slow (or inexistent), international treaties do provide an incentive or benchmark. At the very least, they reveal the extent of local problems. For example:

> The ASEAN[3] and Asian Development Bank initiative on road safety came sometime around 2003–2004[4]; each country within the ASEAN was tasked to develop a national action plan, as well as prepare a country report on road safety. It was only then that I realized how

---

[1]An international treaty linked to the 1992 United Nations Framework Convention on Climate Change (UNFCCC), which sets internationally binding emission reduction targets and commits its state parties to achieve those targets.

[2]Reached at the Paris climate conference (COP21) in 2015, the Paris Agreement is a legally binding global climate deal, which sets out a global action plan to limit global warming to below 2 °C.

[3]The Association of Southeast Asian Nations (ASEAN), an intergovernmental organization, was established in 1967 through the Bangkok Declaration. It aims at promoting regional stability and economic growth among its member states (currently ten): Indonesia, Malaysia, Philippines, Singapore, Thailand, Brunei, Laos, Myanmar, Cambodia, and Vietnam.

[4]The previous ASEAN regional road safety strategy and action plan covered the period 2005–2010. In conjunction with the document, a special, multi-sector working group was established with representation from all ASEAN countries to coordinate and monitor activities at the regional level. The current ASEAN regional road safety strategy was officially adopted in 2015, and further updated in 2016.

> alarming the situation was. … Afterwards, the United Nations proclaimed 2011–2020 as the Decade of Action for Road Safety (Manila).

In response to dire urban transport crises, quite a few current mayors, governors, and presidents have made 'sustainable' transport a centrepiece of their political campaigns, proposing solutions around public and active transport. Even if these are simply political pleasantries, they lead to international searches for models, which can then be used to legitimize new approaches to tackle urban transport problems, (Marsden and Stead 2011):

> If we run into a traffic problem, and we follow our normal way of doing things, we may build more roads. … If we are to formulate and implement a [different approach], this will require evidence of success. … For instance, if we are to stop people in Thailand from using private vehicles, there will be resistance. … That's why we need to use case studies [from abroad] to support policy formulation and implementation (Bangkok).

An interviewee in Malaysia (the most economically advanced among the case studies) underscores that transport 'sustainability' is more about prevention than about applying 'fixes' after transport problems have assumed catastrophic proportions:

> We want to look into how the governments [of other countries] make their citizens adopt electric vehicles and why people choose EVs to invest in. [So far] it's been very difficult to get Malaysians to switch to EVs. It's not like Beijing where they have poor air quality and that has forced their government to create zones for clean vehicles only. But we don't want to wait until when pollution levels in Malaysia become unacceptable; we want to start adopting EVs now. We want to avoid the worst-case scenario (Kuala Lumpur).

Yet, becoming 'sustainable' has proven elusive:

> The word 'sustainability' is attached to every project description… But [knowing] whether [the outcome] is actually sustainable or not requires monitoring and assessment… The problem, I believe, is that there is no follow up procedure to assess whether a project is really meeting the objective of sustainability… We have not seen much [actual] progress. Instead, the number of projects keeps increasing (Bangkok).

A key problem may be that the concept of 'sustainable development' is a sort of Zelig, understood in different ways by various drivers of policy transfer. This is due to the socio-cultural perspective of the latter, but also to the shapeshifting nature—oxymoronic even—of the sustainability concept itself (Redclift 2005):

> The UNDP provided some support for certain projects that were focused on five cities [in the Philippines]. We tried to introduce some good [international] practices. What we found out was these cities were not aware that they were already doing some sort of sustainable transport. They did not know the formal definition … so we had to re-orient them, familiarize them with the concepts (Manila).

Money talks

As expected based on prior research, the motivation for learning from others is often bonded to funding opportunities (Marsden and Stead 2011). However, the availability of donor funding to conduct transport studies does not necessarily lead to implementation, even if it introduces international concepts at policy level. In crude terms, the effect of money on policy transfer—for the better or the worse—is summarized in the following quote:

> The influence of other countries in formulating domestic transport policies is significant when we need funding. If any particular country offers funds, that country pushes for certain policies. If not, we devise our own transport policies. … Some studies are done because the responsible agencies get free money [from donors], and they might as well do a study. These types of studies are seldom implemented; usually they're put away. When an agency really wants to see something happening – probably to comply with national policy or a cabinet order – it will actually put out its own funding to do the study. That's when ideas turn into reality (Bangkok).

If foreign countries or cities—typically in Western Europe or the US in the past, and in East Asia more recently—are willing to provide study funds, they end up being cast as role models:

> In the past, many of our studies on sustainable urban transport were supported by grants and loans from European countries. So, those countries were automatically chosen as benchmarks for Indonesia (Jakarta).

> Which country pours a lot of money here? … Whether they want to or not, the government must accept it. So, we see that now the major players [in Indonesia] are Australia, Japan, and China. [Policy transfer] cannot be separated from politics. I mean, it depends on the orientation of the current administration. If they move closer to China, then the rest must follow. Even if the civil society wants to refer to Colombia, if the government wants to look at China, then all institutions must follow suit (Jakarta).

Besides funding to compile plans and studies, much policy transfer is bounded by the available financing for implementation—whether domestic or foreign. This applies in cases of both voluntary and coercive transfer. Where transfer is voluntary, a local government may be able to choose between several financial models available internationally at one time—such as public, private, or public–private partnership (PPP). But interviewees note that the choice is often not the most beneficial one in the long run, but rather the cheapest in the short term. This is in line with austerity measures and government withdrawal, as promoted by neo-liberalism. At this stage, the neo-liberal ideology is globally diffused and has affected the urban transport sector a great deal (Nguyen and Pojani 2018):

> I am afraid that the private sector will end up obtaining the [new high speed urban rail] concession. But if they do, they'll still need to recruit people from [the public sector] because they do not have the same kind of expertise that we have. The government actually wants the private sector to take care of everything because they do not have money to finance the project… If PPP is the model, the private sector will be in complete control of the project. Is that worth it? Definitely not. In order to maximize the benefit, a better idea might be to look at the [rail development] model implemented in China or Japan, which does not rely on revenue from passenger fees, but can break even and even create profit from TOD [at station areas] (Bangkok).

Again, in accordance with neo-liberalism, certain transport policies which would lead to cleaner air are adopted only when they have the potential to be economically beneficial as well:

> At the national level, support for sustainable transport comes primarily from a competitiveness angle. Environmental concerns seem secondary, … the environmental aspects have just been attached at the end. The issue of carbon credits, for instance, was not discussed at all before. But since the Prime Minister signed an international agreement recently this issue has been emphasised… (Bangkok).

Another aspect of the influence of money in policy transfer is the co-opting of the concept of sustainability by private developers:

> When the government tries to get something done, it will put the label 'sustainability' on the project. Like with the grand development of the Chao Phraya River banks, which only responds to the interests of the private sector, and not people in general (Bangkok).

Recognizing that money talks and always will, cities need to ponder how they can develop solutions that benefit all parties involved.

Image of progress

In the urban branding and policy boosterism era (McCann 2013), Southeast Asian capitals are trying to project an image of success in order to maintain a competitive edge in the region, place themselves on the international map of 'worthy' peers, and leave behind the 'developing city' stigma of the past:

> Thai policymakers like ten-dollar English words that sound cool and are globally known. … Whether they can actually implement those concepts is another matter (Bangkok).

Malaysia, in particular, has very high aims, and benchmarks itself against developed places such as Singapore, rather than less wealthy neighbours:

> There is no doubt Malaysia will be the capital of Southeast Asia in the near future. First, we have the space. We can take over Singapore in the sense of convenience and cultural diversity. But Singapore will still do very well in terms of efficiency (Kuala Lumpur).

Due to the status factor, wealthier cities in the region (Kuala Lumpur and Bangkok) may elect to adopt public transport technologies (such as rail transit) which are above their means:

> What has inspired me a lot is the BRT system[5] in Curitiba and Bogotá. … But of course, the Malaysian public aspires to LRT and MRT.[6] They still don't understand the difference between BRT and conventional buses. There is this prejudice that the bus is not sexy enough! (Kuala Lumpur).

Some planning concepts have been picked up in the Southeast Asian region for the same reason they have diffused elsewhere: they are at a high point in the current planning fashion cycle. To illustrate, bicycling has become popular among youth in Western countries alongside a growing preference for urban (as opposed to suburban) living (Circella et al. 2016), and Southeast Asia is joining the bandwagon:

---

[5]Bus Rapid Transit or BRT is a bus-based mass transit system, which seeks to emulate the performance and amenities of rail-based transit systems. Typically, BRT systems run on segregated lanes. BRT's key advantage over rail transit is its lower cost (Pojani 2014).

[6]Referring to Light Rail Transit (e.g., trams) and Mass Rapid Transit (i.e., metros).

> I think we are merely following global trends... The younger generation sees cycling as fashionable... Bangkok, for instance, keeps trying to emulate initiatives from cities like Bogotá, e.g., the Car-Free Day [Ciclovía] (Bangkok).

The same is true for the notion of urban transport sustainability more broadly:

> Thailand has been aware of the need for, and importance of, sustainable transport for at least fifteen years. ... The reason can be attributed to the global trend of looking at how the transport sector can respond to climate change and resolve traffic problems (Bangkok).

Often, pressures to show action or appear 'modern' overcome actual local needs:

> I think the national government is not thinking rationally about sustainable transport planning. They see that Singapore has a Pan Island Expressway and they too want a Pan Island Link [in Penang]. They want to do something big (Kuala Lumpur).

Even where policy transfer activities appear to be driven by 'logic' and a rational search for better models, at the implementation stage, when it is time to invest scarce public funds on a project, popular pressures take over, and politicians seek to be viewed as proactively making reforms, regardless of how well thought out those may be. Given the progress-oriented culture that reigns contemporary polities, it is easy to conflate all visible investment with positive 'reform'. One interviewee sums up this line of thinking:

> In Bangkok, the effect [of policy transfer] can be seen as early as the planning stage. Public agencies generally employ a theoretical approach borrowed from other countries to construct key performance indicators or the conceptual basis of a project. The question of whether the project will be implemented, however, is not dependent on such information, or the academic perspective for that matter... Decision-making is based on popular support... Instead of identifying shortcomings and then proceeding to collaborate [on solutions], the process is more like, 'people are not happy, let's make changes' (Bangkok).

However, transferring concepts and tools from places with a higher image, and wearing a 'progressive reformer' hat, can be regarded as helpful in improving a city's international standing—and therefore rational:

> Higher-up people have gone on study tours to other countries and seen transport projects that work and are pretty, such as rapid rail systems, and they want to have those in Thailand as well. The real motivation, however, is not transport sustainability, but the chance to display to others that we also have new and modern technology. When you look at it in this light, it makes perfect sense that Bangkok should have rail transit while sidewalks and crosswalks are horrible: non-motorized transport does not showcase modern technology (Bangkok).

> Marikina City[7] is a member of the Alliance for Healthy Cities [an international network initiated by the World Health Organization in 2003]. In 2008, the WHO recognised the Marikina bikeways through a good practice award. ... Now we have to sustain our [bikeway] project because the Alliance members follow up on the awardees. ... Especially when we host an event, we have to showcase our project and provide an update (Manila).

In addition, Southeast Asian cities are coming to realize that a good public transport system can serve as a powerful marketing tool, which can help attract foreign investment and tourism and thus uplift the local economy:

---

[7] A city in Metro Manila.

> Singapore advertises that if you set up a corporate headquarter here, you can comfortably have meetings around the city. Shanghai does the same. They tell the world that if you come to Shanghai, the local metro can take you to meetings in different sites, and you can even go sightseeing in between meetings. … Bangkok is different. Here you cannot plan your [business] activities; you can barely accomplish one meeting per day (Bangkok).

> Having fun transportation modes can attract tourists. It happened in Amsterdam too. Their canals are so attractive as a tourist-oriented transportation mode. Maybe Jakarta cannot implement the same, but the Amsterdam concept was incorporated into a recommendation to the Governor to revitalize ferry transport.[8] So then, we asked the local government to supply the proper boats and they did so recently (Jakarta).

Competitiveness and policy boosterism are also growing between cities within the same country, which act to strengthen the policy transfer drive:

> Right now, [urban governance] is headed toward decentralization. … When that happens, there will be competition between cities. A city will look at another and try to figure out how it could also do this and that. … You can bet that local politicians will want to find something that differentiates them from others, which will positively influence their election campaign and persuade their constituencies (Bangkok).

A desire to emulate places which are perceived as being ahead in the development game stems, in part, from a lingering inferiority complex. Quite a few interviewees frame their country as a 'follower' rather than a leader or innovator in terms of transportation planning policy:

> In our case, a lot of [policy] is driven by the experience of other countries. The Philippines is not at the cutting edge of development. It is playing catch up so it is quite natural for it to devise its policies after looking at policies in other [more developed] places (Manila).

In this context, the backing of an authoritative international organization is often needed to legitimize sustainable transport projects:

> If a proposed policy or tool has not gone through an international forum or a weighty international organization, it is very difficult to gain any traction among decision makers. But if certain policy or idea is pushed through international channels, it is somehow more credible and more likely to be approved… we have more leverage in our proposal… International organizations are considered as the experts… Projects inspired by our own ideas do not get a pass (Bangkok).

In a few cases, resentment or disappointment on having been left behind other Asian peers—not only Western nations—in the 'development' race is evident among interviewees. The following quotes—somewhat tinged by envy—illustrate that the flow of knowledge can change direction, if not easily, depending on which city or country is faster at implementing a planning policy, tool, or infrastructure component (Robinson 2011):

[8] Jakarta's main river is Ciliwung, the natural flow of which was diverted into canals by Dutch colonizers, in the manner of Amsterdam. Despite the canal system, much of Jakarta is prone to flooding, particularly during the monsoon season.

> Ironically, in the 1980s, Malaysians were being trained at the National Center for Transportation Studies [in the Philippines][9] but now they are better than us. Kuala Lumpur really invested. … Bangkok has the same characteristics as Manila but their planning and implementation is good. I will not be surprised if Hanoi overcomes Manila too. Did you know that they are now building a metro – funded by the Asian Development Bank (ADB).[10] … I feel sad that here we've been talking about this since the 1980s and it is still 'under discussion' (Manila).

> Singaporeans, back in the day, used to come to Thailand for study tours of our rapid transit system. But now Singapore has successfully implemented a rail transit system throughout the island… And look at Thailand: our rapid rail transit has barely reached the hundred kilometre mark (Bangkok).

'Work and play' study tours

Since the 1990s, face to face contacts between planners, politicians, and consultants have become easier, owing to cheaper air travel, and rising personal and government incomes in some parts of the Global South. In the past, professionals in the South were 'stuck' at home, and exchanges could only occur when delegations from the North visited. That is no longer the case. Now, South-to-North policy tourism is common too. Nearly all participants in this study make an effort to keep themselves abreast of new developments in the urban transport field by attending national and international conferences, workshops, and other events (as well as by consulting the literature as much as they can). For those dedicated to the planning profession, even private trips become learning opportunities:

> When I travel, I don't just look at [tourist] sights, I also look at traffic and transportation systems. It's in my blood, as a planner. When you look at cities, it triggers your mind (Kuala Lumpur).

> Our previous Mayor Marides Fernando, and her husband, our former Mayor Bayani Fernando,[11] frequented Japan and they took note of the culture of biking over there. They thought that probably one day our city [Marikina][12] could also adopt a bicycle program (Manila).

This attitude helps with the diffusion of policy ideas, although it does not necessarily lead to adoption or implementation. But there are many visits to places of interest which are organized at the institutional level too. In the wealthier cities examined here, including Bangkok, Kuala Lumpur, and even Jakarta (less so Manila), many organizations set aside some budget for study tours or professional trainings abroad. (The typical content of study tours will be discussed later, in Sect. 3.3 *Description of policy transfer*.)

---

[9] A research centre within the University of the Philippines, established in 1976. Previously known as Transport Training Center.

[10] As of 2019, the Hanoi Metro system is still under construction. In addition to ADB, Official Development Assistance (ODA) funding is provided by Japan, China, and France.

[11] The political couple Bayani Fernando and Marides Fernando served as Marikina mayors during 1992–2001 and 2001–2010, respectively (succeeding one another). Bayani Fernando has also served as Chairman of the Metropolitan Manila Development Authority.

[12] A Metro Manila city of about half a million inhabitants, Marikina is now tagged as the bicycle capital of the Philippines (with 77 km of cycling lanes).

While study tours can be rather short and superficial (ranging from a few days to a week each), training sessions can last several weeks and be more in depth. The latter are, of course, more valuable for participants as they can gain a better understanding of context. But shorter tours are beneficial too, as they allow participants to experience or embody planning solutions:

> Representatives from governmental agencies will not be energized by just listening to [academic consultants] preach. We have to make them think on their own. ... We can give presentations and show slides on how to create safe road crossings and a network of pedestrian streets, but that is not as good as letting them experience, absorb, and change their ideas from within. ... More beneficial is taking a group that we want to influence to experience the real thing. ... For example, with BRT outside of Thailand, you can only know what it's like when you ride it. If you spend a day using public transport in Seoul, getting from place to place, you will know how smooth it is, how much time it saves... (Bangkok).

Some interviewees lament the fact that too many policy transfer activities are dominated by 'talking' rather than 'doing' or 'experiencing'. They posit that experiential activities—which can be as simple as riding a bicycle or a bus along a newly built lane—are more effective than presentations and workshops in changing people's viewpoints on transport planning issues:

> Initially, [local planners] did not understand what sustainable transport was and why the roads were congested. There had to be a conceptual shift, and the academic sector ... invited a professor from Vienna to come in and conduct a demonstration. They made a car-size box with wooden sticks and no walls and let a person hold the frame from the inside and walk around to demonstrate how [road] space is used. This showed how road space is wasted and how, if people could be substituted for cars, more could fit into the same space. ... The transfer of knowledge is not about someone coming into tell you what to do. Instead, they have to come in and involve people (Bangkok).

The same eye-opening experience can also happen during domestic study tours:

> We invited representatives from the Medan local government[13] to Jakarta to look closely at the BRT system. Maybe they come to Jakarta frequently [on their own], but they ride around in cars and do not care about TransJakarta.[14] So we took them to the busway, they saw the segregates lanes, the ticketing system... (Jakarta).

In rare cases, a study tour can lead the members of an organization to radically change their concept for a particular project—for the better or the worse:

> Surabaya[15] is partnering with the World Bank to develop a tram system. Initially, they wanted to develop BRT [which is considerably cheaper and quite similar operationally]. ... But along the way, the World Bank invited a delegation from Surabaya to go somewhere. The study tour changed their minds. Now, the city is going to develop rail transport as its backbone. ... Perhaps they realized that's what's more suitable for their city. (Jakarta)

[13]The capital of the North Sumatra province, about 2 million.

[14]Jakarta's BRT system, inaugurated in 2004. Currently covering about 250 km, it is regarded as having had mixed success.

[15]Indonesia's second largest city, after Jakarta.

Besides 'seeing', study tours offer an opportunity to interact with the creators and managers of planning 'best practice', and hear the project's story straight from the horse's mouth:

> [Administrators] have an opportunity to ask questions to the people behind successful transport projects, which they wouldn't have if they just read reports back home. Study tours are not just about going for a visit and looking around. One needs to prepare beforehand and have a goal. ... For example, Koreans are like Thais in many respects, and Seoul is not that different from Bangkok. So why have they found success with BRT, while we have not?[16] ... I believe that if Seoul could do it, Bangkok could also do it. We just need to change certain things. ... Talking to their officials will help us understand how they operate their system, what challenges that they face, and how they overcome them. This is greatly beneficial (Bangkok).
>
> We go [on study tours] to ask about [another city's] experience in implementing the public transport system because this involves not only physical challenges, but also social problems that need to be fixed. We also inquire about issues related to the land acquisition process, the payment scheme, and even the funding arrangements for operation and maintenance (Jakarta).

For cost-effectiveness purposes, most organizations try to cover as many topics as possible within a single study tour, especially when a distant destination is visited. But there is evidence that study tours tend to be more successful when they target a specific issue for which a local organization seeks a solution. A critical attitude is helpful too:

> In preparing for this business [bikeshare], I went to Australia and contacted the people running Melbourne Bike Share. I was more interested in understanding the failures of bike share networks than the reasons for success. At that time, I felt that the literature has many examples for success but there's not much literature on what *not* to do. Melbourne Bike Share launched with a lot of fanfare years ago but it's a notorious case of bikeshare failure globally.[17] So I wanted to know why! [Going to] Melbourne was eye-opening. The study tour made me realise that customer experience and interaction are very important. Very often, infrastructure planners have a mentality that 'if we build it, people will use it'. The trip made us change our business model more than the infrastructure design. ... Now we tell every city in Malaysia that no infrastructure project can succeed without matching educational campaigns and economic incentives. Maybe there should be an increase in parking charges, a congestion tax, a petrol tax? Or things like car-free zones, service subsidies, a free hour of ridership, right? (Kuala Lumpur, emphasis added).

In the case of technical assistance from abroad, it is mutually beneficial if study tours are reciprocated; in other words, if the 'lending' city sends a group of visitors to the 'borrowing' city as well:

> The Japanese team [advising on road safety] sent their students to stay in Thailand with us for couple months in order for them to look around and conduct their own research... I think that the transfer of knowledge in this case was good because we were introduced to the Japanese approach. It was also good for the Japanese students [to come to Thailand] because

---

[16]Up-to-date, Bangkok has created a single BRT line (Wu and Pojani 2016).

[17]Melbourne Bike Share, a government-owned scheme, operated between 2010 and 2019. In its last year of operation, the scheme costs AUD$2 million to run while each bike was only used once a day on average (*The Age* 30 August 2019).

> they could finally study a country that has quite a reputation for dangerous roads. They were able to understand why no result has been achieved even though the problem of road safety has long been known (Bangkok).

However, the interviewees also note that many international study tours or conferences are perceived as paid-for junkets by a portion of participants and therefore any policy learning efforts go to waste:

> One of the challenges that we have, particularly among people with more seniority, is that they do not really pay attention to the lectures and tend to treat the study tour as a vacation. Some even fake illness so as to skip the lectures, but later post on their Facebook profile that they are on vacation… They complain that the program is too long, the shopping time is too scarce. … Once they get on the bus, they move toward the back and start playing cards (Bangkok).

> Capacity building should be real. Not like, you send these people abroad to Korea and Japan, and then you open their Facebook page and you see them posting photos of their trip as if they're on holiday (Manila).

A careless attitude is more pronounced if the study tour destination is too advanced compared to home. In those cases, participants feel like they have 'landed on the moon' and cannot possibly draw any practical lessons; hence they revert to 'party mode':

> Why go and see something so advanced, when you're hardly ready? Just take care of your basic infrastructure first (Bangkok).

Part of the problem is that, in a hierarchical work environment, younger and more passionate planners rarely have a chance to attend study tours abroad:

> Young and enthusiastic officials should be sent to seminars on innovation, but senior executives, who are no longer passionate about their work, tend to be invited to these events instead. … If you see who goes on study tours to Europe, its only the big shots (Bangkok).

The Internet is a small place

The Internet has enabled policy advisors, universities, and governments to retrieve information and publications and/or promote certain practices and areas of expertise much more easily and quickly online. Social media platforms, the use of which may appear frivolous (or even dangerous) to some, have become an important policy transfer tool:

> My friends on Facebook are important sources of information. What they share is news about conferences or projects, along with analysis and discussion. … There are so many websites on urban design and planning that I follow. But I do not use those every day. I primarily go on Facebook to check the links to those websites. … My life now is connected to the internet (Bangkok).

However, the interviewees realize that online planning news come stripped of context. As such, the news may be inspirational but have little practical value, as urban transport professionals need help in understanding the process through which planning organizations go in order to implement ideas, rather than simply marvel at a finished project:

> Nowadays, everybody can search for ideas in the internet but understanding how to go about doing it – the process – is important (Kuala Lumpur).

To go beyond the superficial coverage of planning news on the Internet, some interviewees also subscribe to professional magazines or organizational reports which are also distributed online for the most part:

> I subscribe to and read Railway Gazette Magazine, a monthly publication. … I read the annual reports of rail companies in different countries. Among them, Japan has the best ones. … I also like the reports by Heathrow Airport [rail transit]. The Hong Kong MRT, I think, has a very well-written report, maybe more so than Japan. … Usually, the rail companies listed in the stock market are the ones that have annual reports ready for public access (Bangkok).

In contrast, academic journals—which are also available online[18]—have little value to practitioners who seek to apply lessons from elsewhere:

> Academic journals, they are full of graphs and equations – not that useful when we are doing project work. … They're more for statistics and evidence, but not enough to understand the mechanisms that make sustainable transport possible (Bangkok).

This quote points to a persistent problem in contemporary planning academia, which reaches well beyond transport: its widespread conception as an 'ivory tower' disconnected from the 'real world' and incapable of contributing to the transfer of knowledge from the written page to the urban space (Hurley et al. 2016). However, despite 'the basic tension between the scholarly aspirations of research work and the context-specific drivers of practice', 'exchange between planning researchers and practitioners is essential to the development of both disciplinary knowledge and professional practice' (Hurley et al. 2016, p. 1).

3.2 Drivers of Policy Transfer

> Who engages in policy transfer? What are their roles, attitudes, values, and resources?

A political champion for every project

Political leadership is considered as crucial in all cities in order to achieve transport sustainability:

> If you want to change your country, be a politician. Engineers and planners can only give advice on technical issues, but implementation requires budgeting and goodwill from politicians. [In Jakarta], there are many projects but the implementation rate is very low. Some projects, such as the Mass Rapid Transit are being implemented after decades of planning,

[18]Although most academic publications are behind paywalls, the full text can often be accessed for free through sites such as https://www.ResearchGate.net or https://www.Academia.edu.

> or are still pending. This should be a question mark for all of us. Had the MRT been implemented in the '80s or '90s, Jakarta would not be in this state now. So I feel the bottom line is, top level decision makers, particularly politicians and government officials, must be well educated on what sustainable transport is and what the effects of the policies they have implemented are (Jakarta).

Some interviewees point to the importance of political championship in urban transportation planning and implementation elsewhere:

> What Seoul and Bogotá had was a white knight [referring to former mayors Lee Myung-bak and Enrique Peñalosa, respectively]. We need a mayor like them, a leader with a vision who is not afraid to stick to what should be done. In both cities, BRT initially received much criticism due to the traffic congestion it created. Had that happened in Bangkok, the project would not have continued (Bangkok).

Politicians and planning advocates have a large role to play because, all too often, they are more skilled than academics or planning technicians at marketing ideas to laypersons:

> If [academics] speak in terms of numbers and reports, no one will pay attention. Thai people do not want to read a lot. They want powerful a message. It is an issue of marketing: how do we learn what people want, so that we can turn our research into something digestible and appealing to our target group. I have started to engage more with the media, as well as using communication techniques like exhibitions and concise pitches (Bangkok).

Nearly every major transport project in the region appears to have been the pet of an influential politician, who was inspired by international models. For example, in Jakarta, Governor Sutiyoso championed TransJakarta's creation (Fig. 3.3). In Bangkok, Chamlong Srimuang, the then-Governor of Bangkok, was pivotal in the creation of the first Skytrain line.[19] A controversial riverside promenade plan (including cycling and walking spaces) was also prepared by political fiat rather than careful comparison of international examples, and consideration of potential impacts on the local environment, residents, and businesses:

> The Chao Phraya River Promenade project reflects the culture of public project preparation in Thailand, during which there is no deliberation. It began with the Prime Minister who, while attending the opening ceremony of the Bike Fest, said that he wanted to create bike lanes along the river. That's all he said. Since then bureaucratic agencies are frantically trying to make it happen in order to please their boss. If they're asked whether the project is right or wrong, or good or bad, they'll just say 'I don't know; my boss said so, and I have the duty to make it happen' (Bangkok).

Powerful individuals who advocate for particular sustainable transport concepts are crucial at lower levels of government too—not just at the top:

> In Thailand, sometimes [urban] development is not be about *what* we try to push for, but depends on *who* is willing to struggle to put in the agenda what they deem important (Bangkok, emphasis added).

A preference (or tolerance) for planning by 'strongmen' has its roots in a hierarchical and patriarchal working culture and society:

---

[19]The Green Line, which began operation in 1999.

Fig. 3.3 TransJakarta station and elevated pedestrian bridge

> In Thai culture, senior officials are always right, and junior officials have no right to speak, or can never be right (Bangkok).

This approach can extend to the entire nation:

> [Thai cities] have a history of a [local] political family being very dominant. So if you hear the name of a certain city, you are suddenly reminded of a certain prime minister, whose name is synonymous with that city (Bangkok).

While this local form of nepotism is seen as ensuring long-term commitment to certain projects, if sustainable transport hinges on particular individuals or political families rather than being enshrined in planning documents, there is always a risk that projects may be discontinued if their promoters are overthrown by the electorate, and replaced by politicians who want to diminish their predecessors. For example, one interviewee explains the workings of the Bangkok Metropolitan Administration (BMA):

> There are two parts with BMA: one bureaucratic side and one politically elected side. Part of BMA's planning agenda comes from the governor, who makes promises to the constituents [during the electoral campaign], which will have to be fulfilled after the elections. The political side claims that their planning agenda comes from the people whom they represent… So policy does change with a change in governorship. If there is an election, and a new governor comes in, he will come in with his own policy agenda (Bangkok).

In any case, quite a few interviewees lament that sustainability champions are few and far between. Possibly this is because the socio-environmental benefits of interventions in sustainable transport take years to materialize, and may be less tangible, whereas politicians seek to cram highly visible projects within their mandate:

> The executives here … their heart is not with the issue of sustainability. They are more interested in their own 'sustainability'. They care about how fast a project can materialize without caring whether said project will be sustainable in the long term. … This is your typical Thailand. People think that the quicker a project is accomplished, the better it is. … The ones who care about sustainability do not last long… (Bangkok).

Emancipation of local experts

The more economically advanced cities of region (Malaysia and Thailand) in which the planning sector is becoming more mature, are beginning to wean themselves from foreign technical assistance. A new crop of local experts, often with study stints abroad, is emerging, who is now driving policy transfer:

> I feel like at this point in time we have become knowledgeable and capable of thinking and doing things on our own… We are picking and choosing different characteristics and attributes from many places, and making those compatible with Thailand. [We're] not taking anything lump sum and hope for the best (Bangkok).
>
> I think Malaysia has come to a stage where we are confident enough to find our own solutions for our own problems (Kuala Lumpur).

Local experts are becoming more confident in Indonesia and the Philippines too—though a technological time-lag is still a barrier here, and foreign expertise carries a higher status:

> Our people are actually capable. But we don't have the technology yet. … Maybe we can build a toll road on our own, but if it is something like the Sunda Strait Bridge,[20] that kind of thing needs a technological expertise that we do not possess. … In some cities, foreign consultants are preferred – locals will listen more to outsiders [than to Indonesians] (Jakarta).
>
> What we are finding is that locals are playing a bigger role than before in those masterplan preparation teams… The quality of Filipino consultants today is just as good as the quality of international consultants. So why bother bringing in people who may or may not understand the local situation when there are Filipinos consultants who do understand the context and also have international experience? (Manila).

Technical assistance from abroad is less needed because all countries have achieved a certain level of higher education:

> When it comes to transfer of successful cases, it is not about technique. Those in the field already know this. Transfer in terms of knowledge and education is already here. There are many people with PhDs in transport in Thailand. The problem is how to transfer implementation programs here. (Bangkok).

Those Southeast Asian agencies which are in a position to hire foreign consultants have also come to realize that doing so comes at a cost—if one has to first provide (informal) cultural training for the consultants:

[20] A planned megaproject that will link the Sumatra and Java islands via road and rail.

> I would not hire any foreign consultant. ... In Malaysia, we have dealt with many foreign consultants in the past and we have come realise that we end up spending more time educating them in the local context than they spend providing us with solutions (Kuala Lumpur).

After years of transport planning driven by foreigners, the added value is no longer significant:

> The Japan International Cooperation Agency (JICA) has just finished the Metro Manila Dream Plan.[21] This is the fifth masterplan on sustainable transport in Manila. All five masterplans have been prepared by foreign consultants or a combination of foreign and local consultants. ... If you overlay the five plans on top of one another you pretty much come up with the same plan (except for the most recent one which covers more area because the city has expanded) (Manila).

In some cases, a veritable fatigue with disjointed foreign planning and planners is evident:

> There are tons of studies. ... There have been studies supported by the World Bank, studies by the Japan International Cooperation Agency, studies by the Australian Agency for International Development, studies by the Asian Development Bank. ... How do you consolidate all these studies and plans, all these outputs? So that you can come up with just one plan that you will promote for continuity? (Manila).

In Thailand it is now mandatory that projects which are to be implemented locally be approved by local experts. Financial independence has proven liberating, and is allowing local governments in the region to be more selective during policy transfer activities. Employing locals is also much more cost-effective (in the case of Bangkok, about a quarter of the cost of employing foreign experts). One interviewee explains:

> Thai consultants can do everything now. This is different from the past when, say, the World Bank provided funding for a study and hired German consultants to do the planning. There is nothing like that now in the urban transport sector. If we use Thai money for a project, then we have no need for foreign consultants... Only in the case of new technologies [e.g., high speed trains or integrated ticketing systems], the expertise of foreign consultants may be required for particular aspects (Bangkok).

Reliance on local consultants and confidence in local expertise is positive in that it avoids 'policy gifts' from more advanced lenders, which, experience has shown, often go to waste as locals are unable or unwilling to operationalize them (Pojani and Stead 2018b). It also helps retain wealth in the country. However, local rules may offset the independence of local planners:

> For research services [rendered to government], Thai consultants must be hired. The owners of these consultancies are usually lecturers from [top local universities]... For master plans, there is a rule in place that requires a university to provide the stamp an approval... And that is why companies turn to universities in search of that stamp of approval (Bangkok).

---

[21] Formally entitled 'Roadmap for Transport Infrastructure Development for Metro Manila and its Surrounding Areas', also referred to as 'JICA Dream Plan'. The plan was approved in 2014, and covers the period until 2030.

In addition to increasing local expertise, there may be other, less positive reasons why local companies have come to dominate the market. Besides charging higher fees, international companies may not be willing to give in to local political demands while formulating their analysis:

> Real professional companies [from abroad] cannot survive here, trust me. They cannot compete or maintain their integrity. Those with decision-making power don't really want professionals. Put it this way: when you're deciding where a transit station should be located, a politician can come up and say that they want this station put near their house. If those in power want something, they are going to get it. That's why Thai companies are preferred: they're willing to give in to such demands (Bangkok).

> When the government hires consultants, they have to go through a tendering process, right? So they usually end up with local consultants simply because according to the Indonesian system, the cheapest bid wins. And the cheapest one is usually local, and usually not that good. The BUMDs[22] that run TransJakarta or the MRT can go for good [international] consultants. That's not the case with transportation agencies under the national government (Jakarta).

While local consultants may have gained a stronger position within the respective cities, the voices of indigenous, not-for-profit sustainable transport organizations (another policy transfer driver) remain relatively weak. Manila hosts the headquarters of Clean Air Asia, one of the largest non-governmental organizations in the region (with branches in New Delhi and Beijing), which promotes air quality in Asian cities. While active, the organization is not very large: it employs about thirty people in total. In Thailand, a local initiative called Smart Growth Thailand (SGT) works to promote public transport at the local level.

Overall, grassroots organizations formed around sustainable transport, while well-meaning and somewhat effective, do not necessarily present a united front, nor do they share the same agenda and viewpoints. It is unclear whether they are more effective at managing conflicts than for-profit firms or governmental organizations (Krashinsky 1997). Conflicts occur even when the concept nonprofits seek to transfer is relatively narrow—for example, urban cycling:

> One [local cycling organization] is Thai Cycling for Health Association (TCHA), while another is Thailand Cycling Club (TCC). These two are at odds with one another [laugh]. Together they promote cycling in cities around Thailand. This inspired the Bangkok Metropolitan Administration to turn Bangkok into a cycling city, so the promotion of cycling followed here… But the character of the leadership in these two organisations differs. TCC is led by a retired academic who specializes in environmental management, so it tends to organize conferences and educational activities. TCHA is directed by a cyclist, which explains why it is less interested in promotion and education but rather focused on pushing for exclusive bike lanes. (Bangkok)

The international student diaspora

Many of the local experts who are now acting as policy transfer drivers in Jakarta, Manila, Kuala Lumpur, and Bangkok have trained abroad. The trend to undertake undergraduate and postgraduate studies in other countries—mostly Global

[22]Provincial or municipal corporations, locally known as Badan Usaha Milik Daerah (BUMD).

North ones—is responsible for much policy transfer in the Southeast Asian region. Academia acts as a sort of policy transfer broker. A variety of foreign influences are absorbed through education and training. This trend is reinforcing longstanding legacies because the most desirable study destinations remain Western Europe, the US, and Japan, and academics from those places tend to project their own cities' conditions onto Southeast Asian locales:

> Most of us studied in Japan, and Japanese academics or even practitioners have always in mind examples from their own country. They say that if you flip the map of Metro Manila, it is similar to the map of Tokyo. But we also know the limitations; the things you can do there are different to the things you can do here (Manila).

While students from Southeast Asia may understand the limitations of North–South comparisons, they also internalize a comparative mindset. Southeast Asian academics who train in abroad and return to their home countries[23] unavoidably pass on some of their acquired outlook to their students:

> Where one completed one's doctoral program does have a strong influence on one's outlook… I was trained in America, which sets me apart from those who went to school in Europe or Japan due to different educational approaches [in those places]. For example, those educated in Germany tend to be rather detailed and thorough… (Bangkok).

Some relationships forged with instructors or fellow students during one's studies abroad persevere for years afterwards, contributing to the diffusion of ideas:

> Because I graduated from Japan, my circle is there. Every year my advisor, Professor [X] visits, and conducts a workshop [in Bangkok]. Every time there is a workshop, a seminar is organized too, and local politicians, bureaucrats, and residents are invited to participate (Bangkok).

Meanwhile, practitioners trained overseas may turn into loyal 'customers' of technology and policy produced in the country where their *alma mater* is located:

> The [Bangkok] rapid train system is influenced by the Japanese model. The first plan for a mass transit system was designed by those who graduated from Japan (Bangkok).

Western nations have long understood these outcomes; hence their investments in scholarships for international students from the Global South:

> Because I was educated in Europe, that is where I usually look [for good planning examples]. As I said, [Western] influence through academic training [of foreign students] is a very smart strategy: a low investment with a high return. Graduates of a particular place admire the country in which they conduct their study, and that contributes to the decision to buy technologies or products from that country [when] back in Thailand… (Bangkok).

At this point, national governments in Southeast Asia do not have to rely entirely on scholarships offered by foreign governments, which in the past constituted a form

[23]An academic association, the Eastern Asia Society for Transportation Studies (EASTS), helps local researchers to stay in touch once they return home. It has more than 1,500 members from all over Southeast Asia. It runs a biannual conference, and publishes an online journal (Asian Transport Studies).

of financial aid to poorer countries. They have sufficient internal revenues to award scholarships to local candidates. However, public scholarships to study abroad are more easily accessible by well-connected individuals who belong to the upper social strata. In Bangkok and Jakarta, it is not uncommon for public officials to take time off work to complete PhDs abroad, even if they do not have the intention to join academia in the future. (International scholarship awardees are often required to return to their home city, and serve a number of years in the sponsoring organization.) This situation reflects and strengthens entrenched social hierarchies throughout the region:

> Education is really important [in Thailand] especially for the upper classes… The [Thai] government sends to study abroad those who will have power over the national development. These are university lecturers and bureaucrats. Returning students will eventually play an important role in shaping the national development policy (Bangkok).

While local universities have improved, foreign education imports are regarded as necessary, at least for another generation. According to some views, sustainable transport concepts cannot be borne internally due to a still-underdeveloped local knowledge base:

> Everything in Thailand is based on our study of other countries because we don't have universities that are centuries old… We have to acknowledge that we have not developed our own intellectual basis as adequately as other countries. That's why we have to borrow concepts and tools from them. We need to make a continuous effort to develop our local knowledge because there are many problems in Thailand that require a context-specific approach (Bangkok).

Local knowledge may have been slower to build than expected due to 'brain drain'. Unfortunately, one portion of the youth who study abroad never return—deterred by a perceived institutional dysfunction in their home country (see later):

> A lot of Filipinos work [in Singapore], including some of our former students; they work as [transport] engineers, planners. The people who are building Singapore's subways, a lot of them are Filipino engineers. So you can't say that the knowledge is not there. In fact, some of the students come back to us and say, 'Why can't we have something like this? It's not that hard to do. If only the government were competent.' That's what's sad. If only [the government was willing], there would be no problem attracting people back in the country (Manila).

Technology mongers

Southeast Asian transport technology firms are among the most powerful drivers of policy transfer as they mix and match elements from various countries in their products. For example, the electric scooters produced and sold by a company in Malaysia combine the following:

> Our batteries are from Korea, our technology is from America, some of the body parts are from China, the design is Italian, but the scooters are manufactured here, in Penang (Kuala Lumpur).

In the case of technologies that are not yet present in the region—especially heavy and costly technologies such as rail—overseas producers often send travelling salespersons to Southeast Asian capitals to showcase their products. These salespeople act as policy transfer drivers:

> Foreign vendors want to sell their products and come to pitch the sale. They follow what the trends are (Bangkok).
>
> I guess one of the reasons why TransJakarta is buying Scania buses is because a delegation from Sweden came and they met with the governor and sold their product (Jakarta).

To beat today's rapid technological obsolescence, sometimes cities must purchase new transport technologies directly from vendors, without waiting to see them implemented elsewhere. This has both benefits and risks:

> We are looking at a BRT system that will be operational in five years' time, so if we use one of the existing systems, it will be very backward by then. Technology changes very fast and we are exploring the most advanced options. ... The plan often requires us to look for state-of-the-art technology but the technology must also demonstrate a 10–15 years of track record. That's a bit of an oxymoron. In a sense, it's a matter of whether we're willing to take a risk on [new] technologies (Kuala Lumpur).

Given that local technology buyers have become more sophisticated, foreign technology sellers must compete quite fiercely for business in the region. Technology transfer is being driven by demand rather than supply:

> Honestly, [technical assistance from abroad] is about the vested interests that different countries have. Japan wanted to sell us their rapid train [technology]. Germany had Siemens, which also strove to sell its technology to Thailand. These countries were not in a bidding war, but no one is charitable without wanting something in return. That is why if Japan would send their people here, Germany would do it as well. I am not against Thailand being perceived as a destination for investment. How we will benefit from this opportunity is a more important concern. [In the past], I used to work with both Japanese and German experts. Instead of relying only on one expert, I would ask the Japanese expert first and then test that idea by asking the German expert for comment. They were usually in conflict, and we had to work harder because we needed to take and compare their comments in order to see what would be the best option for Thailand (Bangkok).

Once a particular rail system company has secured a contract in the region, this can ensure ongoing policy and technology transfer activities between its headquarters and Southeast Asia, as experts are sought to provide training on specific aspects of the technology, for example on safety and signalling:

> The decision [on foreign consultant hiring] depends on who we are buying the system from. We will hire the technology's owner. We [usually] have to choose between Europe and Asia.

However, there is some evidence that Southeast Asian cities are attempting to break their dependency on single transport technology providers and switch to open systems that can work equally well with different technological components from a variety of providers. The example below is related to urban rail transport:

> Germany has been very influential because the products we're using come from Siemens... Right now, we are in the process of switching from the old system control, which belongs to Siemens and is closed, to a new one. ... We are looking at the [pan] European model... Europe now has an interoperability policy, which means that any rail system should work with other systems or products. For example, if a train crosses the border from France to Germany, it should be able to continue on the same tracks... We aim for our system to match the Japanese, Chinese, and Thai systems (Bangkok).

Moreover, some Southeast Asian technology producers are becoming competitive internationally and are aiming to sell their products not only to regional neighbours but also to Western cities. Therefore, taking account of the international scene—in terms of research, development, regulations, and market viability—is key to their operations. A representative from a company that is keen to develop and sell electrical vehicles (EVs) in Malaysia explains:

> For EVs, we always look up to Tesla. Of course, Tesla is in a different board game altogether. … We have always been an international company and obviously our cars must be ready for the market overseas. … Within ASEAN, we are the only car manufacturer to develop our own cars from R&D up to production. Other countries in ASEAN only assemble cars without any R&D, so they cannot compete in terms of EV. R&D needs very deep pockets. To produce a new car model, it takes RM300 million. And we always do our R&D according to UK standards, because we produce cars for the international market and the UK regulations are the strictest. We also export to the UK so we keep ourselves up to date with their regulations and tax incentives. The UK gives out a GBP5000 rebate to EV buyers. Malaysia should do this too, we are working with [the government] on this (Kuala Lumpur).

However, as a representative of a new Malaysian company that produces electric scooters notes, the national government does not provide incentives for EV use:

> Malaysia is moving in the right direction in terms of urban transport sustainability but the government is not coming up with enough programs to educate the public and with incentives for people to buy this technology. There are so many 'green' agencies in the government but none of them are doing anything about it. Now only the rich can afford to buy electric scooters from us; the price is quite high for normal people. If only the government gave the public a RM3000 [$700] rebate… Secondly, the government should create different zones within the city: say, allow access to the centre only for private and shared bicycles, public transport, and electric vehicles (cars and scooters). Thirdly, the government should limit [conventional] car circulation by licence plate numbers (Kuala Lumpur).

The quote above suggests that national governments are necessary allies in the transfer of sustainable transport strategies. They need to provide the right incentives to market players.

3.3 Description of Policy Transfer

> What elements of policy are transferred? Is it policies, goals, tools, or programs?

Existing literature suggests that policy borrowers tend to focus their attention on physical development (Pojani and Stead 2015a; Cook 2008; Lee and Hwang 2012; Temenos and McCann 2012). If a picture is worth a million words, a site visit may be worth a million pictures (Fig. 3.4). Seeing is believing:

> One of the easiest ways to communicate with people is through visuals. If their eyes deem what they see as good, it is easy to convince people (Bangkok).

Fig. 3.4 Michelangelo Merisi da Caravaggio, 'The incredulity of Saint Thomas,' c. 1601, Sanssouci Picture Gallery, Postdam, Germany. *Source* Wikipedia Commons/Public domain

The painting depicts the biblical story of Apostle Thomas ('Doubting Thomas') who refused to believe Jesus's resurrection until he could personally see the wounds received on the cross by the resurrected Jesus. In common parlance, a Doubting Thomas is a sceptical or incredulous person who refuses to believe without physical evidence or direct personal experience.

In this study too, interviewees tend to have a stronger interest in foreign *city*scapes than in other aspects of planning. This is understandable given that physical development is visible and comprehensible:

> I think other countries really shape policy in Indonesia, perhaps because of the decision-making style in this country: we must see how something is actually implemented elsewhere before we can adopt it in an Indonesian city (Jakarta).

As noted earlier, viewing and experiencing 'places' is also more enjoyable than sitting in lectures about legal, administrative, and financial planning tools. There is also a belief that 'seeing' is more likely to lead to implementation than 'hearing' lectures on abstract concepts:

> [On study tours] we are interested in smart growth and TOD; we want our municipality people to see actual projects that could be implemented back home (Bangkok).

Fig. 3.5 Singapore's riverside

Popular attractions amongst policy tourists include the re-opened Cheonggyecheon canal in downtown Seoul; the Tokyo metro system; Amsterdam's cycling lanes; London's public bus system; and BRT systems around the world. Aesthetically pleasing projects are particularly inspiring, as humans are naturally attracted to beauty, details, and intricacy (Isaacs 2000):

> The Clarke Quay[24] development in Singapore was the inspiration for Marikina City (Fig. 3.5); that's what the previous Mayor had in mind when he cleared the river. Singapore really was their model. And even the City of Manila wanted to have something like that, especially along the Pasig River area (Manila).

However, foreign projects that are not necessarily 'beautiful' but offer practical and efficient solutions to everyday transport problems—and are, above all, novel (at least to the policy borrowers)—are quite inspiring as well. The following quotes illustrate how local planners have been newly exposed abroad to applications of Park 'n'

[24]A historical quay on the Singapore River, now re-developed to accommodate tourist, entertainment, and commercial activities.

Ride (PnR),[25] solar powered roads, and Universal Design[26]—concepts which have become, or are becoming, normalized in more advanced economies:

> The [study tour group] looked at public transport operations in Germany. In many German [inner] cities, there are no cars allowed. So commuters from the suburbs need to drive or ride a bicycle to a train station, park there, and then ride public transportation. Train tickets are intentionally made cheap so that people do not use cars (Jakarta).
>
> We saw something [different] in South Korea. They created a bike path along the median strip of an expressway. The path is covered along its entire length with a roof of solar panels, which serve a double purpose: [generating electricity and shielding cyclists from the sun].[27] Just imagine if we had such a bike path along NLEx.[28] Imagine if you could bike from Quezon City to Pampanga![29] (Manila).
>
> When we are abroad [on a study tour], it doesn't seem like a 'luxury' at all for disabled people, pregnant women, or senior citizens [to use public transport], because one can easily step [from the vehicles] into the platforms. It seems normal. But here, one needs to climb stairs, there are gaps between the train and the platform... We're used to those kinds of barriers here. So we need to sharpen the sense of our engineers by looking at public transport examples abroad (Jakarta).

Instances of Transit Oriented Development or TOD are another requisite destination of international study tours. TOD is defined as development near, and/or oriented to, mass transit facilities. While the concept is based on nineteenth century ideas of rail-based urban development in industrializing nations (Pojani and Stead 2018a), the modern reincarnation of TOD is much more focused on urban aesthetics than in the past (Pojani and Stead 2015c); hence its attraction to policy tourists.

While in theory and discourse TOD has become fashionable in Southeast Asia, it is a notoriously difficult concept to transfer and adapt across international settings due to its complexity, and differences in cultures regarding spatial uses (Thomas et al. 2018). The following quote confirms that not all local planners are familiar with the core tenets of TOD, and implementation tends to occur ad hoc, when particular individuals with more international exposure are keen on this concept:

> Living in Hong Kong ... allowed me to appreciate the fact that over there urban and residential development go hand in hand (Fig. 3.6). So, for the last 4–5 years, I have tried to influence policy in Thailand [in the direction of TOD]. ... What I did was to get the National Housing Authority to sign memorandum of understanding with the Mass Transit Authority (MTA). ... Finally there is a joint project to design residential land use around the rail network. This is a case directly inspired from a model abroad, but I actually lived in Hong Kong and understood the concept. ... If you ask Thai experts, they cannot say that they truly understand. ... It was my luck that I was convinced that TOD works. ... I also had an opportunity to be involved in strategic planning for the MTA and inserted this agenda item into their plan (Bangkok).

---

[25]PnR facilities allow auto-dependent commuters in the suburbs to leave their automobiles parked alongside rapid transport nodes (train stations, bus rapid transit terminals, and ferry terminals) and complete their journey to the city centre by rapid public transport.

[26]Universal Design is the design of a built environment in such a way that it can be accessed, understood, and used to the greatest extent possible by all people regardless of age, gender, size, or ability (physical or mental).

[27]This refers to the 20-mile expressway that runs between Daejon and Sejong.

[28]The North Luzon Expressway, which connects Metro Manila to the Central Luzon region.

[29]A city and a province north of Manila (northeast and northwest, respectively).

Fig. 3.6 High-rise residential development along rail transit in Hong Kong

Local interpretations of TOD lead to a variety of unexpected problems and contradictions:

> TOD is a typical example of incomplete and selective planning policy transfer… The mass transit authority is supposed to develop the areas around stations or provide incentives to attract private investment. The construction law allows developers of residential high-rises near stations to reduce parking in the hope that people will prefer mass transit over private vehicles. But low-income [and carless] people do not have money to pay for housing [in TODs], and that is why they have to live far away [from the city and from transit stations]. So the law just benefits real estate developers, who do not have to invest in parking space and can then develop more commercial space. Instead of reducing parking, why not increase low-income housing [within TODs], so that more people can access mass transit? So the poor do not have to travel for miles to access rail stations and the city does not have to expand far and wide? (Bangkok)

Issues of TOD gentrification, which the interviewee above highlights, have been noted in the international literature too (for an example from Taipei, see Lin and Yang 2019). Besides TOD and urban design more broadly, even relatively simple technological fixes make a more lasting impression when they are seen or experienced firsthand in other countries—and make people realize what is missing at home:

> The Melbourne road tolling system is very good. It's so seamless that you don't even realise you are entering a toll station. As you're driving, there's no booth, just a small sound 'tick-tock', and your account is charged. So I ask, why does Kuala Lumpur need to build such

a huge structure across the road? Twelve lanes of toll booths in Sungai Besi![30] Everybody needs to go through the booths before converging back to three lanes again! A waste of space, time, money! (Kuala Lumpur).

In Australia, at the train stations, you have an [electronic] board which tells you the trip schedule and all that. You don't have that in KTM[31] stations (Fig. 3.7). The current KTM service is a dinosaur; the management has been resting on their laurels (Kuala Lumpur).

When I was in the UK, I was surprised to see that when a bus was idle, the engine would be automatically turned off. They plug a software into the bus which turns the engine on or off automatically. In Malaysia, if a bus is idle, the engine keeps running. So if buses are stuck

Fig. 3.7 Kuala Lumpur railway station, housed in a historic building (1910) which mixes Western and Oriental architectural design

[30]A highway in Klang Valley, which runs through south-eastern Kuala Lumpur.

[31]Keretapi Tanah Melayu (or Malayan Railways Limited), the main rail operator in Peninsular Malaysia. A publicly owned corporation, it dates back to the British colonial era. In 2001, the Kuala Lumpur railway station's role as the city's main rail hub was supplanted by Kuala Lumpur Sentral (KL Sentral), a modern, intermodal transport hub.

Fig. 3.8 Announced revitalization of the Klang River banks in Kuala Lumpur

> in traffic, imagine how much fuel is wasted! I tried to bring the technology here [from the UK] but it's very expensive. So I'm now looking at Chinese companies (Kuala Lumpur).

Viewing concrete examples serves also to convince planners (and eventually decision-makers) that solutions which did not occur to them or they did not think possible for their cities have in fact worked well elsewhere. That knowledge can then be used to legitimize local proposals:

> In the case of BRT, when the heads of transportation agencies went to Guangzhou and saw for themselves how 300 buses passed a station in a minute, they could finally say, 'Okay, this is really working. This isn't just some number shown to me at a meeting. I counted the buses myself'. … We usually have a plan in mind – before the study tour even. After the tour, it's going to be the same plan, but now people are so much more confident that it will work out. Then it's about the approval. Once [decision-makers] get to experience a project [overseas], we can go for the approval. We lay out the details for them after the study tour – it's not efficient doing so beforehand (Jakarta).

> Many people here argue that we can't do public transport because our streets are too narrow but many European cities have proven that it's possible to fit public transport into narrow streets (Kuala Lumpur).

At other times, a study visit to a foreign city is needed to work out the details of a higher-level plan at home:

> We already know that the general solution [to urban transport problems] is mass transit. But the issue is the details: should the system be elevated or underground, what should the curve radii be in the inner city, what should the distance be between stations, and suchlike. What are the [rail transit design] trends in urban areas with limited land (Jakarta).

In some cases, foreign examples of transport solutions serve as a stimulus for envisioning entirely new approaches—even where these do not exactly match the original.

For example, one interviewee in Kuala Lumpur offers a nuanced view of the local applicability of water taxis along the Klang River. It even touches upon the expectation that public transport be dignified and decorous—a prerequisite in a traditional setting such as Malaysia—while comparing the situation with Asian peers such as Bangkok and Shanghai. In this case, water taxis are a concept picked up in Paris—a city which uses its river only to transport freight and tourist cruises, and has not succeeded in creating a commuter ferry system along the Seine (Bignon and Pojani 2018):

> We are involved in the River of Life (ROL) project (Figure 3.8).[32] I envision that in the future, Kuala Lumpur can have a water taxi system. I learned this from Paris, where of course it is more of a tourist experience. Why can't we have water taxis? Klang River runs through the city centre from north to south. The river is actually wide enough to accommodate water taxis - though not big ferries. Water taxis don't have to be a form of mass transit, we can just use them as a feeder service, at 10 km/hr. This can be a by-product of the ROL project. In Bangkok, they have water taxis plying the Chao Phraya River, and they are so efficient. People don't have to depend on road transport, they just go to the river bank. Of course, the Bangkok system is very primitive, there's no proper anchorage point, so people basically risk their lives jumping into the boat and the boat itself is just a sampan.[33] When it starts off, the water splashes on your face! I actually travelled on those boats once, and I noticed how city employees, smartly dressed - even the ladies in miniskirts - just climbed across into the boat without worrying about, you know, being improper. That's the way of life in Bangkok. … I came up with this concept called 'KL Bund'. I use the word 'bund' as in the Shanghai Bund[34] because it attracts good feedback, it surprises everyone. The KL Bund, as part of the ROL, will make the river part of our life, will turn it into our national icon. … But people have to be realistic about the water quality. It's just not possible to have a river with water as clear as in Pulau Tioman.[35] The Thames in London, the Yarra in Melbourne, the Seine in Paris - even though they don't have floating rubbish, they do smell, their water quality is really not that great. (Kuala Lumpur)

In some cases, those visiting overseas destinations are reminded of historical urban amenities that their own city may have had in the past and has now lost. While dispiriting, this awakening to local assets might inspire a desire to restore more sustainable transport traditions from the past. The following quote, from an interviewee in Bangkok, illustrates this point. It is somewhat ironic as Bangkok was extolled above:

> If we could utilize Bangkok's canals for commuting, then we may be able to address the issue of connectivity (Fig. 3.9). Why do we have to look elsewhere and marvel at how good

[32]A project to rehabilitate the 120 km-long Klang River (Sungai Klang), which flows through Kuala Lumpur and pours into the Straits of Malacca. The river is now heavily polluted and the project seeks to clean the water, while also mitigating floods and developing the riverbanks with commercial, residential, and recreational activities.

[33]A flat-bottomed wooden boat, sometimes including a small shelter on board. From Cantonese sāan báan, literally 'three planks'.

[34]The Bund (Waitan) is a waterfront stretch in central Shanghai. Lined with colonial and other historical buildings, it faces the Pudong District, Shanghai's modern CBD. 'Bund' literally means 'embankment', from Persian, 'band'.

[35]Referring to Tioman Island, an island with pristine waters and lightly developed beaches, situated on the southeast coast of Peninsular Malaysia. It attracts tourists from both Kuala Lumpur and Singapore.

Fig. 3.9 Simple river boats used for transport along the remaining Bangkok canals

> the canals are in South Korea and China? We tend to be so enamoured with all things abroad, like the cruises on the Thames. Our own canals, however, used to be really beautiful. But many are not like that anymore because [the banks] are occupied by squatters or vendors. People cannot really access the areas along the canals now (Bangkok).

While there is an obvious built environment bias during policy transfer activities, many actors in the region are sufficiently sophisticated to scrutinize different 'best practice' examples beyond appearances. Rather than being dazzled, overwhelmed, and discouraged by the achievements of other cities, many look at how those cities were able to proceed from general planning policy to specific projects and finally to implementation:

> Vienna is interesting because it has a high level of integration between policy and practice. … Usually any city will have a plan, but the execution of that plan might not necessarily happen. Vienna, on the other hand, has been able to find interesting ways to make almost all of the items in their plan executable (Bangkok).

> The reason I look at [Copenhagen and Curitiba] is because they have the right vision. They have shown that they not only have the ambition but also the will to do whatever they set out to do. For example, Copenhagen was pretty car-centric once. But starting in '60s and '70s, they converted a lot of car lanes into bicycle lanes and they pushed up their bicycle modal share (Kuala Lumpur).

An interviewee in Metro Manila explains how a local municipality learned from Singapore and South Korea that, to renew an old and polluting public transport fleet, 'carrot' measures for drivers are as, or perhaps more, important than 'stick' or 'control and command' measures. Simply ordering drivers to take action is not sufficient to achieve sustainability goals:

> San Fernando[36] managed to convert its tricycle fleet from two-stroke engines (which are more polluting) to cleaner, four-stroke engines. The city government provided interest-free loans to the tricycle drivers to switch to four stroke engines. ... This was the first ever transport project funded under the ADB's clean technology fund. ... They looked at the experience of Singapore and South Korea. Before, these countries had a lot of bus franchises but they got their operators to merge so that they could have a more effective public transportation system. ... So San Fernando's main strategy was to involve the tricycle driver associations. They focused on the health angle. All tricycle drivers were tested for tuberculosis and other physical ailments. It turned out that many had a lot of lung problems and that's how the city government managed to convince them that there really was a need to renew the fleet. Also, the city government promised that there would be a cap on entry to the tricycle franchise. So San Fernando would not be like other [Southeast Asian] cities where there is such a proliferation of tricycles that on certain days they cannot even operate. ... Here the tricycle drivers have a higher income because there is a cap [on supply] (Manila).

In addition, practices such as public input and participation, which, while 'invisible', have been a planning mainstay in the Global North for decades, are slowing being embraced by Southeast Asian cities:

> Public participation is becoming an important issue when we talk about transportation policy in Jakarta. It's what we've learned when we visited Amsterdam. They said that they had set up a Specific Communication Community during the construction of their Metro. ... However, the [Indonesian] government is not completely open to the idea of citizen involvement in every policy decision; on the other hand, many citizens are ignorant about policy making processes (Jakarta).

3.4 Direction of Policy Transfer

> In what direction are policies transferred? Is the direction intra-national or cross-national?

Diminishing Western role

Large international organizations, such as the World Bank, the United Nations (and its various sub-organizations), the Asian Development Bank (ADB),[37] and the Organization for Economic Cooperation and Development (OECD), are less involved in capital cities in the region than in the past. However, they still provide advice or fund

[36] A medium-sized city located 66 km (41 miles) north of Manila.

[37] Despite its name, 19 out of ABD's 68 members are Western countries outside the Asia-Pacific region. Australia and New Zealand are members as well, and ADB has representative offices in Frankfurt and Washington, DC.

studies on national-level planning issues (such as upgrading fuel standards), or help with the development of impoverished rural areas. An interviewee in Kuala Lumpur recounts:

> When I was still a consultant in the '90s, some of our projects were funded by the World Bank and the Asian Development Bank. At that time, Malaysia needed a lot of help and funding so the country was very receptive of input by the World Bank. Now Malaysia is quite developed. We are sustainable in the sense that we can handle things on our own. The World Bank now funds development projects in rural areas rather than a built-up city like Kuala Lumpur (Kuala Lumpur).

Interviewees in Jakarta offer further insights as to why the World Bank and the Asian Development Bank may have curtailed their involvement in the urban sector in Southeast Asia:

> The problem with the World Bank and ADB is that they want to do everything alone. Indonesia doesn't have the channels yet to drive them into local areas – although that's where things are happening. That's the obstacle, so we want to solve that bottleneck (Jakarta).
>
> I think the World Bank and the Asian Development Bank mostly work on national road projects because those are more appealing than the urban transport sector in terms of economic viability and internal rate of return (Jakarta).

Where city-to-city exchanges occur without the intermediation of large international or supra-national organizations, different Western cities serve as templates for different aspects of planning in Southeast Asia. Policy transfer actors tend to mix and match rather than rely on a single place as a role model.

> I do not necessarily think that we have to rely on a particular country for everything. The success of country X may be a result of its unique context that Thailand may or may not share. We could potentially benefit more from countries with less outstanding models or projects that somehow are a better fit (Bangkok).

Transfer actors borrow selectively from a range of Western places that have come to embody particular policies:

> I tend to prefer European cities. In terms of cycling, I look at the Netherlands. I also like Sydney with respect to the connectivity between mass transit and water transport... Australia is well known for road safety education... In the US, I like Portland, Oregon, which is an excellent case study on TOD (Bangkok).

Seeing as the search for role models is quite dispersed, in most cases it may be more appropriate to talk about 'lesson-drawing' from the West rather than a linear form of 'policy transfer'. It is clear that past colonial relationships, which involved unidirectional transfer between two asymmetrical countries, have now been mostly broken. It is hardly the case that Jakarta partners exclusively with Amsterdam or that Manila models itself after Los Angeles. At times, Western consultants are even viewed with suspicion, given a mixed record of past collaboration results, and their obvious commercial interests. With the Southeast Asian region becoming more sophisticated and discerning, it is no longer the case that any consultants from the West are received with open arms:

> Now are working with Bloomberg and the Rockefeller Foundation… We have signed MoUs with them, but the results vary. … They are selecting cities that can become models for development… and Bangkok was selected … on the basis that it is facing traffic problems. But, because they are from the private sector, we are still not sure if they are going to provide support or want something in return (Bangkok).

An interviewee from Malaysia is more explicit:

> I think that these [Western] foundations have a second motive: they want to get the economy [of their country] going by teaming us up with technology companies from Europe and North America. Green technology can be a good way for them to penetrate our market (Kuala Lumpur).

Meanwhile, some new mentors of a Western bent are emerging, such as Israel. Lacking a strong planning reputation, Israel must offer funding in order to entice locals:

> The Israeli embassy frequently offers funding [for study tours]; the applicant has to be interviewed and selected by them. I have been to Israel on their sponsored trip before. It is quite a rare destination for Thais. Their know-how was very good, and the program took about 15 days to complete. I felt like the thought process was a lot more complex there (Bangkok).

Among the case study capitals, only Manila (the least developed of the four) appears to be more open to Western mentors. The ADB is much more active in Manila, where it is headquartered, than elsewhere, and its assistance is sought after by locals:

> It had been a while since ADB had a project with the Department of Transportation and Communication, and DoTC staff would say, 'why is ADB not helping us'. And I was like, 'probably you should know how it works, you have to propose projects, you have to put it in your [next] Country Partnership Strategy, you should know what's in your [current] Country Partnership Strategy'. The good thing with DoTC is that they are open about the fact that they need support and assistance. … There are some governments that know they don't have capacity but are still very closed. … It is just a little bit off to ask for help on something that they should already be doing internally (Manila).

In the past, large international organizations and development banks have been accused of being deeply ideological, endorsing specific solutions that furthered their own cause (Evans 2009a), or promoting universalistic, one-size-fits-all 'best practice' which rarely worked equally well everywhere (Rose 2004). This study suggests that, in a departure from the coercive practices common in past decades, these organizations are now treating local governments as partners on a more equal basis. The ADB and the World Bank are moving to demand-driven assistance. An academic in the Philippines explains:

> The mandate of international development agencies such as the ADB and the World Bank is to provide assistance… But prior to that, the host government – in this case the Philippines – must already have crafted its own strategy. Then the development agencies will look at the national strategy to identify how they can come in… In other words, assistance or intervention should not be donor-driven or supply-driven. The ADB and the World Bank, they don't come in forcing their agendas, technologies, packages or programs. It's not like that (Manila).

The European 'Pentagon'

While global organizations, in which Europe is a party, are losing ground, the Western European planning 'brand'[38] has been and remains quite influential in Southeast Asia. Southeast Asian transport and planning agencies send many delegates to various cities in Western Europe for study tours. Europe's appeal owes much to its attractive cityscapes in addition to its achievements in sustainable transport. The value of urban aesthetics, culture, and character cannot be underestimated when a country seeks to become a policy exporter:

> I would single out Barcelona. That's a city which has really stuck in my mind. Public transport there is very affordable, hiring a bicycle is easy, the system is so fluid that I did not encounter any kind of difficulties. The history of the city is quite rich, and accessing their different museums and art galleries is always easy. But it's not just their transport system. I liked sampling their restaurants and bars, and talking to the people there. ... Liveability is about the whole package, not just about isolated bits. Transport isn't a standalone item (Kuala Lumpur).

In addition to Spain, cities in France, Germany, Austria, the Netherlands, Belgium, Switzerland, the UK, Sweden, Denmark, and Italy are popular destinations. It is clear that the European core area—the so-called 'Pentagon', delimited by the metropolitan areas of London, Paris, Milan, Munich, and Hamburg (European Commission 1999)—retains a strong reputation as a policy lender. This is a zone that concentrates high-quality infrastructure, wealth production centres, and command functions, and enjoys a privileged status as a 'global economic integration area' (European Commission 1999). More peripheral countries in the European Union, such as those in the Sunbelt (Italy, southern France, Spain) and in Scandinavia (Denmark, Sweden) are influential too, whereas Eastern Europe is missing, nearly entirely, from the interviewees' mental map.

The European core retains its influence in part because consultancy firms therein are active in bidding for transport projects in the region even where the respective governments (including previous colonial powers) have withdrawn their assistance. Some have forged relationships with Southeast Asian partners at a time when government funding was more abundant. Those relationships endure into the present and are revived when new funding becomes available:

> We have a small network of good colleagues in Hungary and Germany, with whom we used to collaborate on a project funded by the European Commission. We completed two phases five years ago, and this year, they contacted me to let me know that there is [new] funding from available from Erasmus, and that we should work together on this [new] project. As with anything, if people know and trust each other, they will continue to work together. But it took time to build that foundation (Bangkok).

This approach differs from Japanese consultants whose links in the region are more often brokered by their government (see later). Communication channels with Europe that have opened decades into the past persist into the present. (For example, in

[38] Brand awareness is a term employed by business to describe the extent to which customers are able to recall or recognize a product. It is a key indicator of a brand's competitive performance in the market.

Bangkok, it was German consultants in the 1970s who planted the idea of rail transit in the city's first masterplan.) However, European governments also make some investment, if only in the form of sponsoring study tours for planners from Southeast Asia, in order to stimulate business for European consultants. One interviewee in Malaysia says:

> Our visit to Copenhagen was actually sponsored by the Embassy of Denmark. Once we've been to a city, we're obliged to consider consultants from that city. There's always a business element in our visits abroad, there are always strings attached. We'll visit their city council, then we'll visit their consultants, service providers, companies, and industries related to [the topic of the study tour] (Kuala Lumpur).

In case of direct, national government-backed technical assistance, only Germany among Western European countries has remained active in Southeast Asian capital cities, through the Corporation for International Cooperation (*Gesellschaft für Internationale Zusammenarbeit* or GIZ). GIZ does not provide funding but limits its activities to advice and awareness-raising. These may sound frivolous but a GIZ representative in Bangkok highlights how much patience and perseverance is needed in shifting current thinking in urban transport:

> The training that we provide does not translate into actual practice overnight. It takes time. ... For example, [participants at a training session] thought that public buses emit more greenhouse gases [than motorcycles] because they're bigger in size... We explained that greenhouse gas emissions from public buses, when divided by the number of passengers, are lower than emissions from motorcycle taxis, which can only carry two people. ... A lot of the participants did not make this connection (Bangkok).

The seemingly simple task of identifying and overcoming language barriers can take years. A GIZ representative says:

> In the first three years, we organized workshops and activities in English but we noticed a problem: participants seemed to understand the content, but did not really comprehend it. So, we switched to Thai and we also worked on the re-design of our content and format (Bangkok).

For obvious reasons, the policies that GIZ promotes—most currently: fuel efficiency, electric two-wheelers, MRV,[39] and green freight and logistics—tend to originate in Europe, and most consultants are European:

> A majority of our consultants are German. ... They do a short visit for one to five days. Because their services are expensive, we don't have them here for long... [GIZ's] concept is to use the services of both local and international consultants, so that exchange can take place... Japanese consultants are very competent, and there are quite a few of them here, but we do not hire them. ... The reason is that we are GIZ (Bangkok).

[39]Measurement, Reporting and Verification (MRV) of greenhouse gas (GHG) mitigation. All Tiger Cub countries are non-annex parties to the United Nations Framework Convention on Climate Change. As such, they are obliged to report on the actions they have taken or plan on taking to implement the Convention. The reporting provisions were further enhanced through the Bali Action Plan adopted in 2007, which introduced the principle of Measurement, Reporting and Verification (UN 2014).

Clearly, there is an element of self-interest in GIZ's work in the region:

> GIZ will tell us, you can do this, this, and that, and then they'll come back and say: look, we have this technology, are you interested [in buying it]? Perhaps, there is a business element after all (Kuala Lumpur).
>
> If the funding comes from Germany, [in theory] consultants do not have to be from Germany; there is an international bidding process. …But coincidentally, German consultants usually win the bid (Jakarta).

Nonetheless, the interviewees appreciate GIZ's small-scale, urban-oriented, and incremental approach which contrasts to the blunter approach of other policy advisors which may not be as effective in reaching local populations:

> GIZ seems to be more tactful. It does small-scale projects. If there are any leads, there's always an opportunity for another small project. It's different from the U.S. which gives out huge block grants. With GIZ, it's more gradual. If there's a gap that's not covered by one project, then it can be translated into an opportunity to create another project. As long as the demand is there (Jakarta).

At the supra-national level, the European Union has had some involvement in urban transport in the region through its Asia Pro Eco project (now more than a decade old). The EU contributed €31.5 million towards the total budget of €82.3 million, and policy transfer was a core task of this project. As the European Commission (2002) press release noted: 'The transfer of know-how and best practice will help build sustainable economic development and environmental management.' At the time, the project led to some transport activities, and some connections between European and Asian professionals have endured into the present.

While European cities continue to be admired, locals also recognize that contextual differences with Southeast Asian cities make policy transfer quite difficult:

> I look at Amsterdam, and that case study is not something that can be transferred here. It is true that a lot of people there bike, but the climate, environment, society, and culture are very different from Bangkok. … We have to look to which Asian cities promote cycling. For example, Tokyo has very chaotic traffic, but it still manages to have bike lanes and people cycling (Bangkok).

Australia as a good neighbour

While the image of Australian planning is not nearly as shiny as that of Western European planning, Australia is generally regarded as friendly neighbouring country which offers some good examples of planning practice in specific areas—for example, zoning regulations, bus rapid transit implementation, road safety, and so on—as well as some generous funding. A preference for visits to Australia owes much to it being physically closer to Southeast Asia than both North America and Europe. It is not uncommon for interviewees to have relatives in Australian cities, whom they visit from time to time. As such, connections with Australia may be 'circumstantial' rather than being driven by Australia's planning reputation. Only Malaysia, as a Commonwealth country, has a specific preference for Australian (and British) planning consultants. In Indonesia, interest in Australian transportation planning is driven by the Indonesia Infrastructure Initiative (IndII), an aid and technical assistance programme funded by the Australian Department of Foreign Affairs and Trade. However, as an interviewee in Jakarta puts it:

> For governance maybe we look at Australia. We do not look at Australia for technology (Jakarta).

Australia is regarded as lagging behind developed Asian countries (i.e., Japan) in urban transport technology—a reasonable observation given limited and obsolete public transport networks (either road- and rail-based) in most Australian cities. Besides assisting Southeast Asian neighbours and attempting to strengthen Australia's influence in the South Pacific region, Australian aid programmes provide work opportunities for Australian consulting firms:

> If the funding is coming from AusAID,[40] normally they also hire experts from Australia (Jakarta).

If Australia were incapable of providing financial aid, it may not have been on the radar as a planning policy lender. As is, Australia's two largest and oldest cities, Sydney and Melbourne, are more often mentioned as study tour destinations. The capital, Canberra, is too small in size (population under 400,000) to serve as an example to any Southeast Asian megacities but it often ends up on the itinerary of official visits by virtue of being the national capital. Brisbane[41] is sometimes visited too, specifically for its BRT system, which has been successful albeit quite limited in size (Chayacani and Pojani 2017).

No role models in North America

While some Western settings (Europe and, to a lesser degree, Australia) remain enticing in Southeast Asia, North America does not emerge as a particularly inspiring role model. Canada is hardly mentioned whereas the United States is notorious as the homeland of urban sprawl and car dependence (although individual planners may be highly capable):

> I will not cite the US [as a place to learn from] because it is so car-oriented. Their cities sprawl far and wide … because they have lots of land… (Manila).

Only certain cities, typically on the West Coast, are mentioned as potential lenders:

> Portland would be that particular city, and the Pearl District [revitalisation] is a model of public-private cooperation which I feel Thailand could follow (Bangkok).
>
> Partly San Francisco because of BART.[42] When we were given a tour, we saw their control centre. I was really amazed by it. They monitor everything real time. … Some people are asking if they can do it here (Manila).

---

[40]The Australian Agency for International Development (AusAID) is a national government agency responsible for managing Australia's overseas aid programme.

[41]A city of 2.3 million, located 900 km north of Sydney.

[42]The Bay Area Rapid Transit, a heavy rail system inaugurated in 1972, which serves the San Francisco Bay Area, California. The total length is 180 km (112 miles).

New York is popular too, mostly owing to its efforts to revive bicycling, whereas cities in the American Midwest are absent from the interviewees' mental map, unless a trip to one of them has been sponsored by a US agency. Dying interest in US planning models parallels dwindling US financial assistance for the urban sector in Southeast Asia. To the extent that the US is involved at the urban level, it is in the context of multi-lateral projects, which are not necessarily dominated by North America. US initiatives in the region do not involve much funding at this point. For example:

> There was an initiative from the USAID[43] called the Energy & Clean Air Project (ECAP).[44] USAID was very interested but my impression was that they just wanted to fund the meetings and not really the hard core studies and projects (Manila).
>
> Right now, we are doing a project on environmental management with the US embassy, through the embassy's network. The project is called Eco-Capitals Forum.[45] The Bangkok Governor signed a 'memorandum of understanding' with the US Embassy, along with thirteen other embassies and two international organizations, including the United Nations. … This is a diplomatic relationship to promote environmental management concepts, not a funding consortium (Bangkok).

Local experts have come to realize that American expertise is useful for certain planning aspects, such as road building technology, but when it comes to sustainable transport they need to look elsewhere given the US's poor domestic track record in this area:

> Experts [from different countries] come with their own specialty. Americans are good with [road] infrastructure, but the Japanese know ITS[46] best, and Europeans have a lot of experience in public transport. … Before deciding which countries to look at, we have to first determine which direction we are headed in. For example, a decade ago, the focus was on building roads, and all eyes were on the US which is internationally renowned for its road-building technology. … But now, we are saying that we no longer want to take that approach. … We've come to realize that we can never build enough road space to satisfy the demand. The old way, therefore, is not the answer, and the government also does not have enough budget nor space to build more roads. The focus has now shifted from 'moving cars' to 'moving people' – and we are emphasising public transport. … It is generally agreed that public transport is not US's strongest suit. … So now we have to look at other examples, countries that have found success with public transport (Bangkok).

---

[43]Stands for United States Agency for International Development, a federal agency primarily responsible for administering foreign aid and socio-economic development assistance. Among official aid agencies, USAID has one of the largest budgets in the world.

[44]ECAP ran between 2004 and 2008.

[45]According to the initiative's Facebook page, 'The Eco-Capitals Forum (ECF) is a global initiative to make diplomacy a vital driver of sustainable cities. ECF serves as a consortium for the diplomatic community to share best practices and challenges, leverage economies of scale to implement renewable energy and waste management solutions, and support and laud efforts by individual embassies to reduce their environmental footprints and costs. As a partnership between the diplomatic community, city government, non-governmental organizations (NGOs) and local businesses, chapters of the ECF provides a unique platform for communities to come together for sustainability.'.

[46]The abbreviation stands for Intelligent Transportation Systems, an umbrella term which denotes the application of information and communication technologies in the field of transport.

Even the Philippines, a former US colony,[47] has come to recognize that following in the footsteps of the metropole has generated more problems than it has solved:

> Years ago, our government just adopted models used by the Americans without considering the context. For example, when the North Luzon Expressway[48] and the South Luzon Expressway[49] were developed, the forecasting methods were of course based on American standards without even calibrating them to our local environment and behaviour. What happened? Within five years, the highways became congested because we did not consider the types and modes of transport. The Americans only have three: rail, bus, and car – whereas we have up to ten: rail, bus, mini-bus, jeepneys,[50] AUV,[51] tricycles, pedicabs, habal-habal[52]… In the Philippines, we are blessed with a set of public transport modes that offer a type of door-to-door service not found in developed countries (Fig. 3.10). … Unfortunately, because

Fig. 3.10 Jeepney in Manila

[47] Between 1899 and 1946.

[48] Built in 1965. Formerly known as the North Diversion Road and Manila North Expressway. It connects Metro Manila to the Central Luzon region.

[49] Built in 1969. Formerly known as the South Superhighway, Manila South Diversion Road, and Manila South Expressway. It connects Metro Manila to the Calabarzon region.

[50] A popular means of public transportation, indigenous to the Philippines (Mateo-Babiano 2016). Jeepney's colourful and kitsch decorations are considered as a symbol of local culture and artistic expression. To improve passenger comfort and reduce emissions, a jeepney modernization programme was launched in 2017.

[51] Stands for Asian Utility Vehicle—a cheaper, smaller, and simpler version of a Sports Utility Vehicle (SUV), marketed to developing countries in Asia. Popular models in the Philippines are produced by Mitsubishi and Isuzu.

[52] Any two-wheeled motorcycle, ridden beyond its passenger capacity.

> of our training, usually our models are [still] from the US but now we tend to adjust based on our own situation (Manila).

This comment points to the responsibility of policy lenders to take a prudent approach when suggesting or promoting their own models to other countries. However, using today's perspective and hindsight to judge American pro-highway policy advice issued to developing countries decades ago may be unfair. In the 1950s and 1960s, freeways were thought to be capable of alleviating traffic congestion inside cities, decreasing the number of traffic injuries, increasing people's access to jobs and services, facilitating the movement of goods, reducing vehicle operation costs, fuel consumption and tyre wear, and showcasing a city's modernization and economic development. Now, as planning thought has advanced, and societal preferences have shifted, a number of American cities are beginning to remove their ageing highways—albeit very gradually (Khalaj et al. 2020).

New Asian mentors

As the influence of Western nations wanes, wealthy Asian neighbours are replacing them. East Asian policy lenders are not new in Southeast Asia but have recently amplified their participation. Southeast Asian planners are obviously pleased that emerging economies in their backyard can finally serve as development guides, especially as the Western role models of old are perceived as declining:

> Asian emerging economies are growing rapidly as opposed to Europe which is quite stagnant (Kuala Lumpur).
>
> The economic slowdown in [North] America and Europe is a factor that has shifted our focus of cooperation to East Asia (Jakarta).

Japan, Singapore, South Korea, Hong Kong, Taiwan, and to a lesser extent, mainland China come up regularly in interviews. In other words, the original Tiger Economies are already serving as models to less developed Asian countries. South Korea has already instituted KOICA (Korea International Cooperation Agency),[53] which is active especially in the Philippines and Indonesia. But among Asian policy lenders, Japan is the undisputed leader—praised for its superior planning, workmanship, and engineering:

> If we are talking about role models, Japan tops the list. Whatever we do, we refer to Japan first because they have been so successful with their initiatives. Japan is the only developed country in Asia that has 'best practices' for us to learn from, and not just on sustainable transport (Bangkok).

Singapore follows close behind:

> Singapore has relevance here in Kuala Lumpur. Like us, it started as a village. … We have witnessed firsthand how Singapore transformed and progressed (Kuala Lumpur).

The interviewees are aware of Singapore's trajectory from a 'typical' Southeast Asian city to a Tiger Economy with an advanced transportation system. This is the reason for much admiration:

---

[53]Established in 1991, and named after JICA (Japan International Cooperation Agency).

Fig. 3.11 Metro station in Singapore. Users are gently instructed to stand in line by the doors

> We know that we have limited resources but Singapore basically started like that too, with limited resources. … But look at the constraints and look at how they did it. Probably it can be an inspiration for our cities as well. Slowly. They did not become high tech overnight; did not start out having a subway already (Fig. 3.11). They had to struggle through that. Even though some say that they did it fast, if you talk to those who were actually involved in the implementation, they'll tell you that they had a hard time. It also took them time to establish [pedestrian] standards and guidelines, for example to widen sidewalks. Eventually, they persuaded people to take public transportation – buses and subways (Manila).

A representative from a city in Metro Manila, Marikina, reports:

> There is no need for a visa to go to Singapore. … So we adopted Singapore as our international benchmark during the term of Mayor Marides Fernando.[54] In fact our vision then was to be a little Singapore (Manila).

Exchanges with Singapore are facilitated through membership in the Association of Southeast Asian Nations (ASEAN). At the same time, Singapore is somewhat regarded as an outlier given that it is an island city-state which implies no physical space to expand but also no need for metropolitan coordination, and tight immigration controls. Moreover, some policy transfer actors realize that they also need to draw lessons from countries in Asia that are just a few steps ahead rather always looking at

[54]Maria Lourdes 'Marides' Carlos-Fernando served as the Mayor of Marikina from 2001 to 2010.

much more advanced (and therefore incomparable) countries such as Japan or even Singapore:

> South Korea is the next destination because it is considered as a country in transition and has encountered similar problems to what we are facing. How they overcame such problems and made progress in sustainable development is something we can learn from. This is quite different from Japan, which does not seem to have any problem (Bangkok).
>
> We Indonesians have all gone to Singapore. So, a lot of the time we need to get [decision-makers] to think of somewhere else besides Singapore. For example, when we discuss pedestrian planning, many people will say, 'We need Orchard Road[55] like in Singapore'. But, there isn't just Singapore out there. Like I said earlier, maybe they need to visit Bukit Bintang[56] in Kuala Lumpur actually. Maybe that's more relatable – and better than Singapore (Jakarta).

At a very basic level, it is easy and relatively cheap for planners in Southeast Asia to travel to other Asian countries rather than flying to another continent. Hence, more advanced Asian neighbours are used as models even for relatively simple features such as bicycle parking:

> I am interested in bicycle parking space because we do not have that. I went to Kyoto to attend a seminar… I did learn a lot about the bicycle parking system near their metro, which was nicely built and had plenty of slots with security guards (Bangkok).

It may be that countries like South Korea and especially Japan—being physically and linguistically less accessible to Southeast Asians than Singapore, Hong Kong, or Taiwan—have succeeded in promoting the more successful aspects of their domestic urban transport systems, while managing to conceal problems. Hence, they have positioned themselves as 'ideals':

> Of course, we are not really aware of the problems faced by South Korea or Japan in implementing their projects. Maybe there are things we do not know because we cannot penetrate the news and we rely on English translations (Manila).

Existing studies (set in Western Europe) have made a similar observation: when engaging in policy transfer, lenders from government agencies are often more interested in enhancing their own reputations than in helping others. Hence, they highlight their stronger urban programs, activities, and policies, while dismissing or minimizing the impact and cost of unsuccessful initiatives (Wolman and Page 2002).

Even if the manner in which information is 'sent' by 'policy lenders' and/or is 'received' by 'policy borrowers' is distorted (Wolman et al. 1994), other Asian countries are attractive due to a shared ethnic background, which some policy transfer actors find comforting. One interviewee says, only half-jokingly:

> Japan is a key role model for me; it could be because when I walk the streets, I see black hair just like mine (Bangkok).

---

[55] An upscale boulevard, lined with shopping malls and hotels, spanning about 2 km across Singapore's centre.

[56] A glamorous shopping and entertainment district in Kuala Lumpur.

Cultural similarities are crucial too in preferring to be guided by other Asians (rather than Europeans or North Americans):

> We are looking at Asia because the … mindset, culture, and people are more familiar to us. [Asians] prefer flexibly over strict adherence to rules, have a more compromising attitude… When we look at Europe, it is very different (Bangkok).

However, East and Southeast Asia is hardly a cultural block. Many cultural nuances are encountered throughout this broad, populous region. The following quote illustrates differences amongst nations, which can assist or hamper policy transfer:

> Japan does not have undisciplined citizens like us [in Thailand]. It is very orderly and thus easy to manage. But Koreans are similar to us, very chaotic. … I was just in Suwon where a neighbourhood [Haenggung-dong] participated in a month-long car-free campaign [in 2013]. When we got a chance to talk to our counterparts there, we learned that it took three years for officials to convince the locals. There were those who agreed and those who disagreed with this policy, and there was protest. But in the end, social pressures forced dissenters to finally turn around. In Japan, by contrast, when bike lanes were introduced, people did not oppose them (Bangkok).

Another reason for a growing interest in other Asian countries is an increasing consumption of cultural products made elsewhere in the continent, in conjunction with the 'Western diet' of Hollywood movies and Netflix TV series. These awaken people's curiosity about neighbours and their urban systems. An academic in Thailand offers:

> I also try to look at [planning] examples in Asia, especially South Korea, because students have an interest in that country as they have seen it on TV. For example, the re-opened Cheonggyecheon canal [in Seoul] and the recreational spaces for walking and exercising around it are featured in some Korean TV series (Bangkok).

As for direct, government-to-government technical assistance, Japan is the most powerful policy lender in the region. The Japan International Cooperation Agency (JICA) is involved in myriad urban transport projects in all the case study cities. Japan appears to follow the same self-serving pattern of assistance previously modelled by European and American policy lenders:

> In the last ten years, the Japanese have been the main group of foreign consultants. Japan came [to Thailand] with a lot of investment money, and JICA provides loans to Japanese investors. If we use Japanese money for a project, then we have to utilize the services of Japanese consultants; this is stipulated in the contract. … There is nothing free in this world (Bangkok).

A JICA representative in Thailand elaborates on Japan's aid strategy further:

> It is Japanese government's policy to get Japanese companies more involved abroad. We try to motivate Japanese companies to be involved in projects in Thailand. We are not just handing out money but also acting like a catalyst between the public agencies on the Thai side and Japanese contractors. … When you build a new railway line, it is quite risky for the private sector to bear all the construction cost, as well as the operation cost. … You should make a public investment first, and gradually give out concessions to [private companies]. … If JICA funds an infrastructure project, that project will have more credibility in the eyes of Japanese companies. … If Japanese companies join the project, Japanese know-how will

> also be transferred to the Thai side. … Japan's strengths in the transport sector are on safety, comfort, reliability, better product life-cycle … for both hardware and software (Bangkok).

But rather than regarding this as a subtle form of exploitation, locals accept it as the ordinary course in human relationships:

> These organizations [JICA and ADB] usually send experts to work with us, and yes, regarding certain issues, they try to propose policies that benefit *them*. But we can't deny that when we provide assistance to [poorer] countries like Laos, we also do the same thing (Bangkok, emphasis added).

Self-interest notwithstanding, there is strong evidence that Japan's commitment to the Southeast Asian region is long-term rather than transient and superficial. In a number of cases, Japanese experts have taken the time to understand the local context, train local peers, help customize policies, and see technical assistance projects through to completion. The following quote illustrates one such example:

> We went on a study tour to Japan to see how land readjustment was done there, and there was also a study that went on for a long time with Japanese experts coming into assume temporary residency with us. That led to the creation of a new law, agency, and fund for this particular issue. The resultant law is implemented in actual land readjustment cases and is influenced by the Japanese model (Bangkok)

Due to the longevity of relationships between Japan and Southeast Asia, and the reputation of Japanese work ethics, locals prefer Japanese advisors to others:

> Our current President's policy to turn to China is a little bit tricky. … Japan is a proven [mentor], and it is the biggest lender. … And they have experience working with us. Many of their projects have been brought to completion. As Asians, we know the Japanese work ethics, so Japan is a better match for us. … It's a matter of culture, knowledge, and how much we trust them… (Jakarta).

Moreover, funding from Japan appears to be more generous than what is offered by Western mentors:

> GIZ organizes workshops, but we need to have some matching budget to contribute to the organization of these workshops, whereas JICA is not concerned about that (Bangkok).
>
> The World Bank and the Asian Development Bank run capacity-building programs but most of those are concentrated in the Philippines. Local staff receives the training here. JICA and KOICA, on the other hand, they usually offer one month training for Filipino staff in Japan and Korea (Manila).

However, JICA loans are more common than grants in more developed countries such as Thailand, and matching local funding is usually required at the implementation stage. Japanese assistance is also quite structured and cooperative rather than ad hoc and coercive:

> In the case of Japanese experts, there is a letter coming in every year, listing different projects and asking whether the department would be interested in doing these. The letter will also indicate what kind of experts the Japanese side has, what support they could lend, and what we could contribute. If we are interested, we can commence the exchange (Bangkok).

The approach described above differs from the short-term European and especially American missions to developing countries throughout the 1990s, which were one-sided and pushed for the adoption of cookie-cutter planning laws and policies everywhere. Moreover, Japanese mentors appear to be more thorough; they follow up with local organizations beyond the technical assistance stage:

> The results of this [environmental management] program are very concrete, and the City of Fukuoka, [Japan], has consistently sent their staff to follow up on how we have implemented their program (Bangkok).

On a personal level, Southeast Asia is an attractive destination for Japanese advisors given its warm weather, beautiful beaches, interesting historical sites, and hospitality culture:

> We have visitors from Japan who would like to vacation in the Philippines; I also invite them to lecture in the School (Manila).

In the past, Japanese interests may have not always been aligned with the interests of lenders from elsewhere, or with sustainability needs in the region. One interviewee implies that pressures from the Japanese car industry, aiming to increase car sales in Southeast Asia, delayed for decades the development of a rail transit system in Bangkok:

> There is gossip in transport circles that a plan for a rail system completed by German consultants [in the 1970s] never came to fruition because of Toyota… Eventually, mass rapid transit came here, but it took forty years for it to happen. (Bangkok)

However, that is already in the past. Now, Asian-born sustainable transport concepts, ideas, and technologies tend to be of Japanese origin. But China is also building a profile as a powerful transport technology vendor in the region, due to its growing production of bus and rail rolling stock. As a newly emerging economy, it is more willing than Japan to meet Southeast Asian neighbours halfway, as an interviewee in Indonesia notes:

> For the high-speed rail, we know from the news how intense the competition was between China and Japan. China was ultimately selected because according to the government and the Ministry of State Owned Enterprises, if I'm not mistaken, the requirements put forward by China were more relaxed for the government of Indonesia (Jakarta).

China has also gained some reputation as a policy learning destination, because, under pressure to alleviate urban air pollution, it has made an effort to organize meetings in order to workshop potential solutions. Southeast Asian peers have been invited to attend, thus strengthening China's standing as a host. Guangzhou's BRT system—the largest in Asia to date—is also a highlight of many study tours. Its reputation has been enhanced not only by its quality but also by the promotional efforts of the Institute of Transport Development Policy (ITDP), the non-profit organization that helped create it:

> When we were looking at putting BRT along the federal highway, we went to Guangzhou to look at BRT's technical aspects. Back then, we were supported by ITDP in the US; they actually recommended the Guangzhou system to us (Kuala Lumpur).

Finally, around 2010, China put itself on the map through a unique technology it invented, the Transit Elevated Bus (TEB),[57] which proved to be 'sticky' (even if unviable) due to the novelty factor, and was covered extensively in international media:

> We are now looking at China for TEB. TEB is creating a storm, because it's new, people want to know about it (Kuala Lumpur).

However, China's reputation as a policy lender is still weak relative to Japan or even South Korea because it is seen as a place which has yet to solve its own urban transport problems. The proverb 'physician, heal thyself' comes to mind here:

> I'm not sure if we look at China [as a role model]. I guess there are some worthy examples like BRT or even bikeways. But we see that China is having serious problems with regards to motorization. Whereas South Korea and Japan – even though they are heavily motorized, they also have very good public transport systems. That's why we refer to them the most (Manila).

Similar to Japan, the US, and Europe earlier, China has a self-serving agenda when engaging in policy lending activities:

> Typically, these countries [Japan and especially China] really want us to use their technology, labour, or even raw materials. This is a way for them to market their own products, so we have to make sure that the technology can actually be transferred, that our workforce can be involved, and raw materials can be supplied domestically... (Jakarta).

Rare South–South exchanges

Beyond other Asian countries—be they wealthy or not—other places in the Global South are rarely considered. The words of an interviewee in Malaysia sum it up:

> We definitely won't organise study tours to Africa or the Middle East. The City Council (DBKL) usually go to Europe, Australia, and the US for study tours – and now increasingly to East Asia (Kuala Lumpur).

There is some evidence of snobbery, in that only places which are regarded as being 'better' are deemed worthy of consideration—and this eliminates much of the Global South from the list:

> I have never been to Africa or South America. I had an opportunity but declined because my condition is that I have to go in order to learn something and bring it back to develop Thailand. That's why I have to go somewhere that is better, not worse, than Thailand. ... Possibly, Seoul is a city that inspires us because its scale and motorisation level are similar to Bangkok. But Seoul is better than Bangkok, so we can use it as a model. ... (Bangkok).

Exceptionally, some very specific borrowing, most often centred on the concept of BRT, is taking place between Southeast Asia and Latin America. In part, this owes to the efforts of charismatic Latin American politicians—mayors like Enrique Peñalosa of Bogotá and Jaime Lerner of Curitiba—who have been successful not only at creating BRT at home but also at promoting their city's achievements abroad. The interviewees testify to this:

[57]A guided bus which hovers above road traffic. Also called 'straddler bus' or 'tunnel bus'. The concept became popular in the mid-2010s but was later scrapped due to safety and other concerns.

> I like Latin American cities, although I have never been there and have only read about them… I attended a UN meeting in Bangkok on BRT, and the mayor of a Latin American city was invited to speak… Bogotá and Curitiba, these cities share similarities with Bangkok; they used to be really bad, experienced more problems from urbanization than us. But they were able to resolve those problems through transport (Bangkok).

In addition, the role of a 'Western' organization, the Institute for Transportation and Development Policy (ITDP), has been instrumental in intermediating relationships and policy transfers between Latin American cities and Southeast Asian cities interested in BRT. ITDP is a relatively small non-profit organization, which focuses, *intra alia*, on promoting and developing Bus Rapid Transit systems around the world. The organization, which was founded and is still headquartered in New York, was never coercive, and its regional offices in the Global South employ mostly local staff. Those offices are in touch with one another, thus circulating the BRT concept around the globe, and ITDP's American origins help lend credibility to its proposals. For example, one representative in the Jakarta branch says:

> ITDP had to go directly to the governor, promoting TransJakarta and educating everybody about what's happening [with BRT around the world]. That wasn't going to be as effective if it were coming from TransJakarta itself. … ITDP has been doing this work globally for thirty years now. On the design aspect, we [ITDP Jakarta] received a lot of help from ITDP China which has really, really deep technical expertise. The promotional and educational approaches on the larger scale, those are developed by ITDP New York.

While ITDP's typical approach is to collaborate closely with local organizations, in some cases it goes directly to international organizations in order to push certain projects forward. For example, a representative in the Jakarta office says:

> We like working with the Asian Development Bank because, with them, our recommendations will be implemented. Because they are loan bank, they look for projects to fund.

In a few cases, ITDP have effectively used successful examples of sustainable transport interventions in other cities in the Global South to 'shame' Southeast Asian planners and decision-makers into taking action in their own city:

> We took Indonesian government representatives to Guangzhou [to visit the BRT system]. But they were very sceptical. 'Can we do this?' 'Indonesia is different to China.' So ITDP held a seminar and we invited experts from various countries: from ITDP Africa, ITDP USA, all the ITDP regional branches. When locals were shown the BRT systems in Africa, they said, 'Wow, African cities are less developed than us. If it can be done there, why not here'. …. It inspired them, and they became enthusiastic (Jakarta).

While most interviewees have a penchant for Global North examples, some recognize that it is important to look at the achievements of places which are at a similar or even lesser level of development, rather than always 'reaching for the stars':

> In terms of ranking, Bogotá is considered much worse than Bangkok. It is plagued with problems like the local mafia and high political barriers. Bogotá, therefore, is a good case that illustrates how important political leadership is. … If, on the other hand, we say to ourselves that we want to be like Tokyo, everybody will listen, but when it comes time to plan and implement, nobody will act, and there will be many excuses as to why that cannot happen here (Bangkok).

Funds for study tours to distant places, where available, are reserved for 'the usual suspects'—in other words, wealthy cities, whether in Europe, the US, Australia, or elsewhere in Asia. While Bogotá and Curitiba are mentioned time and again, rarely have the interviewees visited them in person due to the distance and lack of support for travel to those places. Hence knowledge of the Latin American context and approaches is somewhat superficial. Bogotá or Curitiba serve more as inspirational icons or signs than as practical examples that can be deconstructed into specific components:

> The BRT system in Bogotá is a very good example... But I don't know how they have so much road space that could be allocated for an exclusive BRT lane. Although Bangkok replicated BRT, ours was more of a pilot project and not a comprehensive network (Bangkok).

Bogotá's BRT system, TransMilenio, appears to be resting on the positive reputation it acquired shortly after its launch in 2000. The interviewees seem unaware of the crisis which the system has been experiencing more recently (Gilbert 2008; Gómez Rodríguez et al. 2017) (Fig. 3.12). Other commentators have noticed that the reputations of certain so-called best practices snowball simply because observers become self-referential (Marsden and Stead 2011). Particular experiences, tools, or places are then cast as templates or icons (Cochrane 2012). In fact, upon closer observation, many purported urban success stories may amount to little more than myth (Wolman et al. 1994).

Beyond Bogotá and Curitiba, other Global South cities crop up once in a while during the research but the mentions are fleeting and the exchanges impressionistic, attained during a conference or other short visit:

> Recently, I was in Istanbul with a group of colleagues for a conference on transport, (CODATU). I was impressed with their trams, light rail transit, and BRT. Their BRT had

Fig. 3.12 Public transport riders queuing in Bogotá. A red TransMilenio bus is visible in the background

> a very high headway[58] and a lot of riders, and the trams penetrated right into the market (Bangkok).

Although this type of exposure may not have a long-lasting effect by itself, if combined with other information that a planner or politician may have accumulated it can help reinforce the idea that other Global South cities are taking sizable steps to address their transport issues. This can create a sense of urgency that can perhaps lead to action in one's home city.

Apart from borrowing the policies of others, Malaysia, as the more developed among the places studied here, is emerging as a policy lender to countries farther down the development leader (in particular India, perhaps owing to the links of Indian migrants to their home country). As such, Malaysia is experiencing some of the same barriers faced by traditional (i.e., Western) policy donors in the past (e.g., corruption and clientelism in the borrowing country, and the like). But, as a country which was until recently considered as 'developing', it may be in a better position to cope with barriers:

> We, along with all the highway concessionaires in Malaysia were invited to provide advice to India and Pakistan on their national highway systems. They asked us to apply our success story with the North-South Expressway[59] … and the road projects that have been developed in Kuala Lumpur. Make those happen in Pakistan and India. In those two countries everything is political, there isn't much [rational] thinking. I find that we are more mature in terms of transportation planning and implementation. Corruption here is bad but they are even worse (Kuala Lumpur).

However, exchanges with other Global South partners are regarded with some amount of suspicion and prejudice, while Western or Japanese partners appear to enjoy more trust in Southeast Asia. One interviewee notes that some Asian companies are prepared to lower product quality in order to cut costs:

> We always work with Indian and Chinese companies cautiously because when they come up with new technologies, they're quite good, prices are very cheap, and everyone is happy. After a few rounds though, they become complacent and start slashing costs by lowering the quality of their products. So we are very careful about that (Kuala Lumpur).

Local hierarchy and rivalry

All case study cities are in communication through membership in ASEAN, a formal organization which serves, *inter alia*, as a platform for sharing best practices in transport and other areas. In 2008, an ASEAN Environmentally Sustainable Cities (ESC) Award was established as a way of recognizing exemplary initiatives. Moreover, the ASEAN transport ministers have adopted the ASEAN Transport Strategic Plan 2016–2025,[60] which outlines numerous goals and milestones to support

---

[58] Meaning bus frequency.

[59] A 772 km expressway which runs from the Malaysia–Thailand border in the north to the Malaysia–Singapore border in the south, connecting all the major cities in western Peninsular Malaysia.

[60] Also known as the Kuala Lumpur Transport Strategic Plan. This plan was preceded by the Brunei Action Plan (BAP), adopted in 2010.

sustainable transport (i.e., low-carbon modes, energy efficiency, user-friendliness, transport and land-use integration).[61] These exchanges take place because (at least a portion of) local planners and decision-makers—especially in Indonesia and the Philippines which are less wealthy than Malaysia and Thailand—understand that there is considerable advantage at learning from neighbours with a similar level of development:

> In the case of MRT or LRT, we do not have [planning and design] experience. So we look at existing projects, for example in ASEAN. European cities are not at our level… We look at places which have similar characteristics to us. That's the point (Jakarta).
>
> For BRT, we look at Jakarta and Bangkok. … We've been surpassed by other countries. BRT should have been done a long time ago over here. … Of course in Bangkok they are still faced with traffic problems but they've been successful in expanding their mass transit despite the turmoil in their political system (Manila).
>
> In Kuala Lumpur, you can see that we've built one elevated walkway from the Pavilion [a mega shopping mall] to the KLCC [the high-rise city centre], and there will be four more. … Partly, we've built the walkway with Hong Kong's and Bangkok's elevated walkways in mind. Bangkok has been quite creative in recent years. They've built the hard structure for their [elevated] light rail transit system and instead of just bare pylons going up, they have created a separate level for pedestrians only, separated from car traffic. By minimally increasing the cost, they've added walking space in the sky (Kuala Lumpur).

European-based studies have similarly found that local governments tend to have a strong bias towards neighbours, which are perceived both as trusted peers or respected competitors with similar circumstances (Wolman and Page 2002). Also, there is rivalry at work among neighbours, and a desire to catch up:

> I guess, the pride also [plays a role in policy transfer]? Sometimes [local planners and decision-makers] find out that Kuala Lumpur or other Malaysian cities have achieved something, and then they're like, 'we cannot lose out to them'. It happens quite a lot that when we have meetings with the Jakarta government, we hear things like, 'Malaysia has already implements this, so why don't we also do something like that' (Jakarta).
>
> Our attitude is: if they can do it in Singapore, Vietnam, Thailand, or other countries in [ASEAN], why can't we do it here? (Manila).

Some places are motivated to make their urban transport sector more sustainable, not only to benefit their residents but also in order to gain primacy and recognition within the region:

> For EV (electric vehicles), we always have to look beyond Southeast Asia because in ASEAN, Malaysia is the most active country embarking on this technology. We want to be the ASEAN EV hub. (Kuala Lumpur)

Capital cities as torchbearers

Jakarta, Kuala Lumpur, Manila, and Bangkok are the definite focus of the respective nations. The following quote from an interviewee in Jakarta summarizes the primacy of capital cities:

[61] However, an interviewee in Jakarta reports that, until now, there have been no concrete actions based on this plan.

> The provincial media talk more about Jakarta than their own cities (Jakarta).

While capitals are larger and more complex than other cities within the respective countries, they also have more resources to tackle transport problems. Costly infrastructure, in particular, is first tested in capitals before being deployed to smaller cities. Once people experience the benefits of a sustainable transport initiative in the capital, groups emerge to push for the same initiatives in the provinces:

> With the Blue and Green [rail] lines now operating in Bangkok, people have learned that mass transit can alleviate traffic problems, and provide a shorter and more convenient commute. Hence the idea of sustainable transport has expanded to other provinces (Bangkok).

As such, capitals serve as the blueprint for development in other cities. This increases the capitals' status but leadership implies responsibilities too:

> I think Jakarta is the trendsetter [in the nation]. So, if policymakers want to implement something in Jakarta, they have to be extra careful as it could be replicated by other cities. And it's quite possible that the decision making and transfer processes in other cities will not be as well thought out as in Jakarta (Jakarta)

On the positive side, smaller cities can avoid newcomer costs by learning from the shortcomings of certain transport projects first implemented in the capitals:

> We are working on creating a full BRT system in Medan. But we don't want to repeat the same mistakes of TransJakarta – like designing a 'closed trunk system' without feeder services.[62] We have to go one step further [than Jakarta]. Medan can still learn from Jakarta by emulating the good bits and making better decisions (Jakarta).

But sometimes, when smaller cities regard the capital as an outlier within the nation, they may end up following lesser role models which are ultimately inappropriate:

> Many cities in Indonesia are building a so-called BRT. But they're not copying TransJakarta, what they're really copying is Trans Jogja.[63] When they see TransJakarta, they think 'oh, it's Jakarta', 'we don't have as much money as Jakarta', 'we don't have such big roads' etc. So, they don't think it's relevant. But sadly, Trans Jogja is not really BRT. It's a bus system run by the government. It has stations but they're on the sidewalks. It doesn't have exclusive busways. … [Trans Jogja was only created because] the national government donated thirty buses to the local government, which the local government didn't know what to do with… Those buses created a problem [rather than solving one]. It was basically a waste of money. … But since Trans Jogja was launched, and the national transportation agency saw it, they were like 'see, if cities like Jogja can do it, maybe this concept can be implemented throughout Indonesia.' So we're seeing all these Trans Musi,[64] Trans Metro Pekanbaru,[65] Trans Sarbagita,[66] etc.[67] (Jakarta)

---

[62] TransJakarta launched a feeder system in 2011, seven years after the beginning of operations.

[63] Yogyakarta's BRT system, operational since 2008. Yogyakarta (often called Jogja) has 4 million inhabitants in its metropolitan area.

[64] Palembang's BRT system.

[65] Pekanbaru's BRT system.

[66] Denpasar's BRT system. Denpasar is the capital of Bali (an island and a separate province).

[67] *The Jakarta Post* (10 October 2019) recently announced that five smaller Indonesian cities—Bandung (West Java), Batam (Riau Islands), Makassar (South Sulawesi), Pekanbaru (Riau), and Semarang (Central Java)—will receive 21 million Euro (Rp 326.5 billion) in grants from the governments of Germany, Switzerland, and the United Kingdom to develop BRT systems.

While a willingness and openness to learning from more advanced capital cities within the same nation is positive, smaller cities in the region also realize that they are not scaled down versions of megacities. As such, they may require different transport strategies in order to achieve sustainability at a reasonable cost (Pojani and Stead 2015b):

> A lot of other cities in Malaysia are looking locally at our big city – Kuala Lumpur – for inspiration. As they embark on a path toward better public transport, they've come to realise that light rapid transit (LRT) is too much for small cities. In cities like Ipoh, Melaka, or Penang, LRT would be overkill. So they're looking at other modes, active mobility, introducing more pedestrian paths and cycling paths, improving bus service, hoping to pave the way for trams, BRT, and monorail (Kuala Lumpur).

In a few cases smaller cities have outdone capitals—to the discomfiture of the latter:

> Isn't it a shame that Cebu City[68] should be the first in the nation to implement BRT[69] when Metro Manila has compiled a BRT pre-feasibility study since 2007, with USAID funds…[70] Apparently, the Mayor of Cebu and his very pro-active, very smart planning officer made a trip to Curitiba … and they said, 'we should also have a BRT system in Cebu'. The planning officer had contacts in the World Bank and other funding organizations; these organizations were impressed with the local activism and so it was the Public Private Infrastructure Advisory Facility[71] that offered assistance… A South Korean firm won the bid for the detailed design (Manila).

Global circuits of policy mobility?

Many countries and cities cropped up in the foregoing discussion on the direction of policy transfer. On the face of it, policy ideas may seem to be circulating in all sorts of different directions, as posited by policy mobility theory, and the directionality cannot be traced or mapped (Robinson 2011). Some of the data in the present study support this position. For example, a common stance among the interviewees is:

> We don't really care where the ideas come from; as long as they're good, we'll adopt them. … When you benchmark, you don't benchmark a single city (Kuala Lumpur).

In part, an impression that multiple circuits of knowledge coexist derives from the fact that international 'best practice' examples are routinely employed to justify any plan, project, or decision:

> It is very rare to see any new idea, which is rooted in our own foundation. Every time, when there is a presentation, whether on urban planning or transport, a case study from abroad is referenced (Bangkok).

---

[68]The Philippines' second city, with nearly 3 million inhabitants in the metropolitan area.

[69]As of 2019, the Cebu BRT system is not operational yet.

[70]The Manila BRT project, one of 75 infrastructure flagship projects promoted by President Rodrigo Duterte's administration, is now moving forward, with technical assistance from ITDP China and financial assistance from the World Bank, the Asian Development Bank, and the French Development Agency (Agence Française de Développement or AFD).

[71]A multi-donor knowledge-transfer organization administered by the World Bank.

However, upon closer scrutiny, it is evident that policies still tend to circulate in the richer-to-poorer direction and not vice versa. Most of the time, a country or city needs to acquire wealth and power before it can serve as a reliable policy lender to others. The following quote is suggestive:

> You see the guardrails in Singapore – they look exactly like the ones in Kuala Lumpur. That's what I have observed but it doesn't mean that Singapore follows Kuala Lumpur or Kuala Lumpur follows Singapore. I don't know! Most likely, though, it's Kuala Lumpur following Singapore, right?! (Kuala Lumpur).

Figure 3.13 illustrates the directionality of transfer more explicitly. Overall – with few exceptions—it is poorer cities, which are lower in the global urban hierarchy, that seek to learn from wealthier counterparts. Zooming in, an unspoken hierarchy based on level of wealth and development is evident among Tiger Cubs too, which conditions the direction of transfer: Kuala Lumpur > Bangkok > Jakarta > Manila. While Manila looks up to the other three capitals, and there are some exchanges between Jakarta and Bangkok, it is not the case that Kuala Lumpur is willing to learn from Manila.

Surprisingly perhaps, the map in Fig. 3.13 looks remarkably similar to the hierarchical system of 'world cities' formulated as far back as the 1980s—before the end of the Cold War (Friedmann 1986). According to that hypothetical system, 'world cities' are those highly 'central' locales which contain global control functions, concentrate and accumulate international capital, and are subject to large migration, both domestic and international (Friedmann 1986).

The three traditional command centres of the global economy—New York, London, and Tokyo (Sassen 2001)—are still on the map. Large planning consultancies, such as AECOM, may have regional branches in the Southeast Asian region but their headquarters remain in command-and-control in Western cities (e.g., Los Angeles in the case of AECOM). However, there is more emphasis on Tokyo (as well as Seoul and major Chinese cities) in this case given that the vantage point of the study is Southeast Asia. While the richer-to-poorer direction of transfer still holds, the difference with the past is that the rich core is now moving away from the West and towards East Asia (whereas Australia's influence is 'circumstantial', as noted).

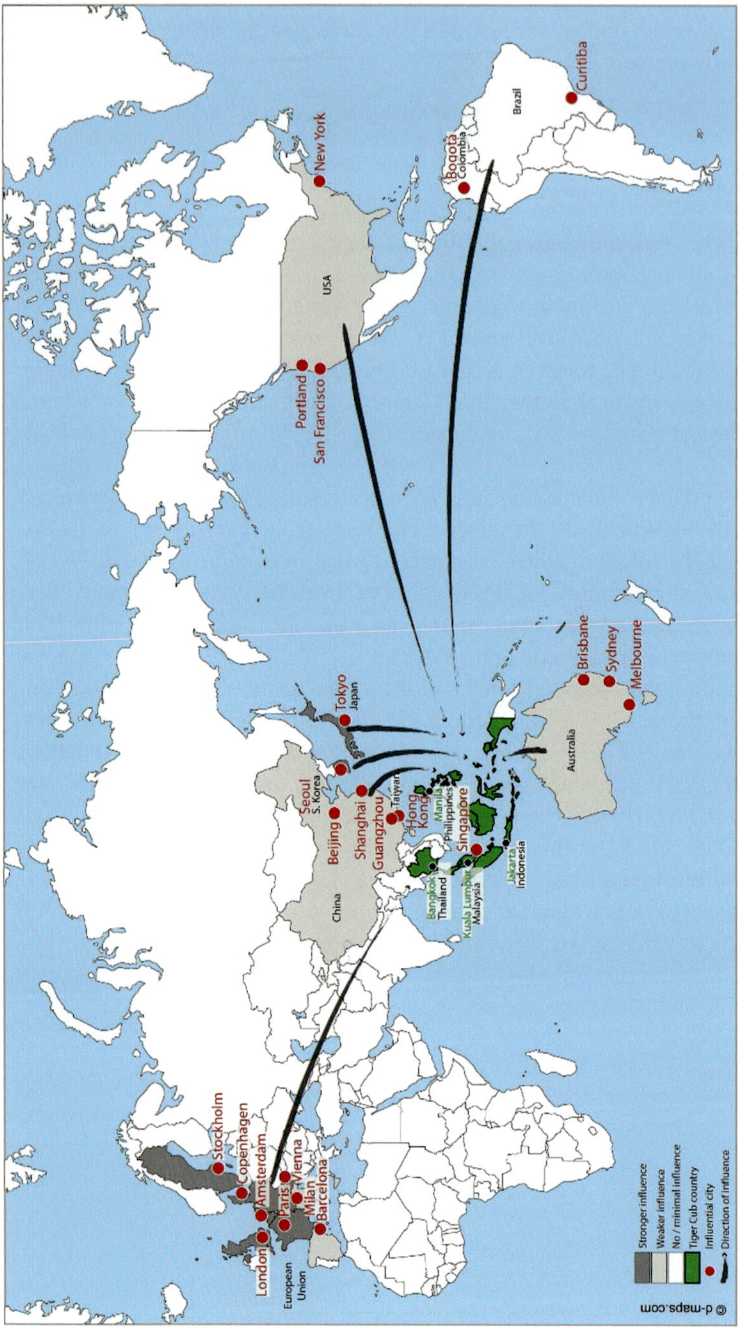

Fig. 3.13 Direction of policy transfer to the region. Map by author based on https://www.d-maps.com

3.5 Deterrents to Policy Transfer

What constraints does transfer face? Are barriers cognitive, structural, or technical/technological?

Money talks

Money was listed earlier among the drives for policy transfer, but local budget limitations are a major deterrent. The following quotes highlight the perceived financial barriers to planning policy transfer.

> From my perspective as a practitioner, knowledge is one thing; actual implementation in this country is another thing altogether because of budget limitations (Bangkok).
>
> It is just frustrating that, since many of us are trained abroad, we usually incorporate a lot of international ideas in the plans we develop, but these plans are shelved. There is no budget or political will from the top to implement our plans (Manila).
>
> I think the most important issue in transport projects is *financial* sustainability. It's not just about environmental and social sustainability. The financial aspect will decide whether we continue with a project or we do something else (Kuala Lumpur, emphasis added).

In Indonesia, the ongoing decentralization reform, which was purported to abate corruption and strengthen democracy, has made urban transport financing even more difficult than before:

> Regional autonomy regulations have made it difficult for the national government to transfer funds to local governments. Meanwhile, local governments cannot possibly fund [large transport projects] on their own (Jakarta).

Besides physical infrastructure costs, complex urban transport projects present high personnel costs, which are unaffordable to understaffed Southeast Asian public agencies:

> A recent urban project that we are interested in is the integration of the bus system in Seoul. We understand that Seoul started from a similar situation [to ours]: many many routes had to be re-planned, redesigned, rearranged, and it is only in the last years that they could reap the fruits of their labour. The current staff of our Land Transportation and Franchising Regulatory Board[72] is hardly capable of matching the level of effort that was required in Seoul. I understand that 300 engineers worked on that project and the LTFRB doesn't even have 300 employees in total. ... I think the engineers are there but the system to absorb them is not there (Manila).

High costs make transport 'sustainability' seem like an optional luxury at times:

> I think the priority for large cities in Indonesia right now is [transport] availability. There are stages: availability is first, and then safety, comfort, timeliness. Sustainability is not a

[72]A national agency under the Department of Transportation, which is in charge of policies, laws, and regulations related to public transport services (including urban public transport).

> priority at this point. When availability and safety have been achieved, then we may start thinking about having an environmentally-friendly transportation system (Jakarta).

> Only those cities which are already well beyond the provision of basic needs can really think about [transport] sustainability for the long term. Once you've fed and housed the people, you can worry about how to move them gracefully around the city (Manila).

Due to lower budgets in the borrowing country than in the origin country, the quality of physical items such as infrastructure may be diluted en route during transfer:

> The first thing that we have to consider is that Thailand is a middle-income country, so our budget is limited. We have to see what we can do with limited resources. We cannot do what rich countries do. … But you have to pay to get something good. People here are not rich, so you have to lower the quality of your product. … Everyone wants high speed [urban] rail, but the question is: how much money do we have? Inexpensive construction cost is one of the reasons that underpinned our interest in BRT (Bangkok).

Public transport remains unprofitable in all cities under study, a fact that is difficult to accept by those in charge of public finances:

> Another obstacle to sustainable transport is the perspective of the Ministry of Finance, which can be described as profit-oriented. The Ministry does not like the fact that [rail transport] is always in the red. … But profit cannot be the exclusive aim [of public transport]. For example, if all bus lines were privatized [and charged market fares], people would not be able to afford to ride (Bangkok).

To recover at least some of the cost of expensive rail projects, some Southeast Asian cities are forced to charge high fares which hurt those with limited financial means, reduce ridership, and lead to duplicate, inefficient services:

> Rail fares in Chinese cities are up to four times lower than in Bangkok. Here, even if you are a professional, using the Skytrain every day really drains your income. … Ordinary people are then forced to use buses and get stuck in traffic… For this reason the bus routes that run underneath the elevated rail corridor cannot be removed… (Bangkok).

Financial limits are an obvious consideration in urban transport planning everywhere. However, in the case of Southeast Asian capitals, one wonders whether the real reason for incomplete or even failed policy transfer is a lack of financial resources or particular beliefs about the legitimacy of subsidizing public transport. Governments here have substantial resources to devote to improving transportation in their capital cities—much more so than many other nations in the Global South. The amounts spent on building and expanding roads (and urban rail systems in some cases) prove this point. Moreover, in all four cities citizens spend substantial amounts on private cars and motorcycles. As an international advisor in the Philippines underscores:

> [Policy transfer] is not only about the money. Remember that the Philippines has evolved in terms of its fiscal capacities. … So from a fiscal standpoint it has a lot of money, it is highly liquid. It does not have to borrow from [the ADB] to build roads, bridges, or any infrastructure… (Manila)

Institutional gaps, overlaps, and turf wars

The institutional framework related to urban transportation planning is crucial, as this is what drives what actually gets accomplished. Institutional gaps and overlaps are commonplace in all four capitals. Urban transport issues fall under the purview of a plethora of public agencies, city officials, national government officials, and other stakeholders whose interests are not necessarily aligned. In Kuala Lumpur the complex local institutional setup stands in particularly high relief because locals tend to compare it to the integrated and streamlined approach of nearby Singapore:

> Unlike LTA[73] in Singapore, which controls both public transport and roads, the jurisdiction of our transport system is split between SPAD, JKR, DBKL, KTM, etc. Cycling and walking are under DBKL; local roads are under JKR; highways are controlled by various concessionaires; KTM Komuter is under the jurisdiction of KTM; buses, BRT, MRT, LRT, and monorail (and in the future high speed rail) are under SPAD.[74] Previously, before SPAD came in [in 2010], the setup was even more disjointed and complicated.[75]

In Jakarta, a policy advisor from an international organization laments:

> In the past, we had leading ministries. Now governments are based on a coalition,[76] so party affiliations affect the cooperation between ministries. This is unlike other ASEAN countries – where sustainable transport is handled by their transport ministries. The Ministry of Transport leads. If they want to do [a project on fuel efficiency], the Ministry of Energy must follow, the Ministry of Environment must provide support, and the Ministry of Industry must make the necessary adjustments related, for example, to vehicle specifications. That's not the case in Indonesia. As development workers, in order to influence [urban transport] policy, we need to approach more than one ministry. We must go here and there. In Malaysia, we would only need to approach their Ministry of Transport and they would inform the rest. In Indonesia, our main partner for fuel efficiency is the Ministry of Transport, but we also need to allocate resources to build partnerships with the Ministry of Industry. When we analysed their strategic plan, we found no mention of fuel efficiency. We first need to make sure whether this is included in their key performance indicators or not. If not, then they will not do it. First, they won't care. Second, they won't have the budget to do it. Third, even if the project moves forward, they won't be able to claim it as their achievement. So, we have to be familiar with their plans (Jakarta).

While the interviewee above praises other ASEAN cities and countries, the present research suggests that in Bangkok, the institutional setup is as, or even more, convoluted than in Jakarta. More than a dozen agencies are involved in urban transport, which report to three different ministries as well as to the prime minister's office. The local council (Bangkok Metropolitan Administration) is but one of many official actors that provide and manage transport within the city. Local planners are well aware that this is neither efficient, nor unavoidable:

---

[73]Land Transport Authority.

[74]SPAD (Suruhanjaya Pengangkutan Awam Darat) is also known as the Land Public Transport Commission. JKR (Jabatan Kerja Raya) is the Malaysian Public Works Department. DBKL (Dewan Bandaraya Kuala Lumpur) is the Kuala Lumpur City Council. KTP (Keretapi Tanah Melayu) is the main rail operator in Peninsular Malaysia. KTM Komuter is a commuter rail system.

[75]In 2018, SPAD was dissolved, rebranded as Agensi Pengangkutan Awam Darat (APAD)—meaning Land Public Transport Agency—and absorbed into the Ministry of Transport.

[76]During the New Order era (1966–1998) under President Suharto, only three political parties were allowed to exist in Indonesia. Now, the number of parties that participate in elections has increased substantially (38 in the 2009 election), which precludes a majority winner; therefore, coalition governments have become the norm.

> Even different units within the Ministry of Transport have difficulty talking to each other. The Minister really has to force them to talk and cooperate, and if that's the case, let's not even mention the level of coordination between the Ministry of Transport and the Bangkok Metropolitan Administration… In contrast, the New York Subway is under the direct control of the city (Bangkok).

No single agency seems to possess the hierarchical line of authority in Bangkok. Function overlaps are significant, yet any collaboration between divisions requires the formal approval of several committees—but also their goodwill (Wu and Pojani 2016). One interviewee highlights the lack of coordination during policy transfer efforts in the area of traffic safety:

> The World Bank has provided financial assistance [to the Bangkok Metropolitan Administration] for educational campaigns… One of the campaigns is on encouraging children to wear [motorcycle] helmets… The impact of these campaigns… comes down to the traffic police actually enforcing the rules. No matter how many campaigns we [BMA] do, if the police doesn't arrest those who break the law, then it's no use. … The traffic police must be transferred to the BMA because urban management has to be the responsibility of the city government, not other agencies, or else there is a conflict of authority (Bangkok).

The urban road network is managed by different agencies, including the Bangkok Metropolitan Administration and the Department of Highways. The ownership of the rail system is fragmented too, with different lines built and overseen by different agencies, which makes coordination rather difficult. This is what an interviewee recounts:

> Although the Airport Rail Link (ARL) stations have lots of space, they don't have many commercial vendors and coffee stands… ARL is trying to manage the stations, but the State Railway of Thailand (SRT) refuses… ARL cannot manage the parking spots beneath the stations either, and has been requesting SRT's approval to do so. SRT, however, remains silent, neither refusing, nor acknowledging the request (Bangkok).

The institutional arrangements in Manila are even more tortuous:

> You always try to get best practice done but then it depends on the local institutions; how can you push them. … I guess the key thing about Metro Manila is the complete and utter lack of coordination or body mandated to do any kind of coordination on transport basically. That's what it really boils down to. The Metro Manila Development Authority (MMDA) really has no mandate and, even if they did, the local government units cannot be forced to do anything, so it's really a mess. … The planning is by the Department of Transportation and Communication (DOTC) but they really don't do any integrated plan for Manila. … The urban planners aggregate the sectoral plans and they really don't have enough independence to be able to question whether or not the plans are viable and if they are actually integrated between different modes and effective. Railways put up their thing and buses put up their thing (Manila).

In Bangkok and Jakarta, nearly all comprehensive urban transport studies—the first of which were prepared four or five decades ago—have remained on paper. Instead, stand-alone mega-projects and incremental planning have dominated the scene. In Manila, higher level visions for comprehensive plans cannot be articulated into operational action plans:

> I think under the Philippine Development Plan,[77] they probably made some noise about 'sustainable transport' or at least 'environmental friendliness', 'green growth', and such concepts, which are also related to transport. But, while there is a high level mandate, coming down from that you immediately hit institutions which are incapable of delivering the aspirations [of the national development plan] (Manila).

Any urban transport elements which require cooperation at the metropolitan level—even technology-driven ones such as traffic control centres—cannot be implemented:

> Davao's traffic control center is fantastic. ... It works just like in Singapore. ... Why do they have a traffic control center in Davao[78] and not in Metro Manila? I think we know the answer. It's because they have a very powerful mayor who has total control. In Metro Manila, the responsibility lies across 17 mayors... Which of these is going to agree or take responsibility for this one? In a place the size of Metro Manila, with this many people in charge, it's very difficult to reach consensus and implement anything [at the regional level]. ... I think there should be a Metro Manila governor – an elected official with power over the mayors (Manila).

The Bangkok interviewees report that, often multiple agencies work in parallel on the same plan or project; then, only one version is selected, while the work of the other agencies goes to waste. Individual agencies charged with executing transportation plans tend to cling to their own historic plans, agendas, missions, and priorities, as they do not have a duty to comply with interagency plans unless instructed to do so by a higher authority (Wu and Pojani 2016). In Jakarta, relationships within individual transport agencies are strained too:

> Things inside the Ministry of Transport itself aren't solid. Sometimes the Director General and the Secretary General aren't on the same page. Sometimes ... the Minister says one thing but the official under the Minister says a different thing. Even different directorates under one Director General don't necessary have a unified voice. It's confusing. That is how the Ministry of Transport works; there are many cases like this (Jakarta).

If a higher-level institution, such as a ministry, has working relationships with international organizations such as the World Bank or the UNDP, that institution tends to 'hoard' and 'monopolize' those relationships rather than inviting urban-level organizations to share information. Moreover, each agency prepares its budget separately and according to a different timeline. This is most ineffective in the case of major transport projects, which need inter-sectorial input. A Japanese advisor in Bangkok offers this perspective on institutional barriers in Thailand:

> In Japan, we have a national development plan, a very broad one, and then a prefecture plan, then a municipality plan. Each plan is related. So, the municipality plan is based on the national development plan. Here, it seems like the NESDC [Office of the National Economic and Social Development Council] has a very broad plan, but I do not see any relation with [lower-level] plans. That makes regional development along the high speed rail difficult (Bangkok).

Beyond desktop planning, the following quote illustrates how detrimental turf wars between transport stakeholders can be on the ground, when trying to 'import' innovative tools:

---

[77] The current plan covers the period between 2017 and 2022.

[78] A city of 2.5 million inhabitants in southern Philippines. It has the largest land area in the country.

> Traffic police believe that they understand traffic better than engineers. When the engineers were setting up an automated traffic light system, the police put up a fight arguing that they could direct the traffic better than an automated system (Bangkok).

The case above can be chalked up to Luddism or technophobia among traffic police officers, who fear redundancy and replacement by technological devices. However, other instances of conflict are reported in which one agency can scuttle the plans of another:

> Even though the [Bangkok] BRT has traffic markers, cars can still get through, and if the traffic is too congested, the police allow cars to go into the BRT lane. I really don't know how to explain this; it's definitely not like this in Seoul or Bogotá (Bangkok).

Strong regional or metropolitan planning agencies, with overall decision-making power and dedicated funding, are seen as the solution here:

> Cities have to get serious about institutions for implementing [mass transit initiatives]. And they can get serious if they so wish. And when people do get serious, stuff does happen. Finally the metro in Jakarta is getting going. Delhi did it; it was a horrible mess and finally they got to a point that they had to do something. The Delhi Metro Rail Corporation was created, and given a mandate, and somebody was in-charge, I think he was a general, so it got built. At the very least, transit authorities should be at the metro level; preferably they should be development corporations. The problem is that, due to the fragmented nature of Southeast Asian metropolitan areas – lots of different cities, lots of different districts within the cities, very limited revenue-raising powers – transit authorities cannot work unless they are directly funded by the national government. Or you can have some kind of corporate entity that can cross-subsidize within itself like the Hong Kong [value capture] model.[79] In most of Southeast Asia, local government revenues cannot fund public transport. Local governments can't even manage to free up their footpaths, they can't even build footpaths, let alone build mass transit in any way, shape or form. That needs a different level of organization (Manila).

Even with the current lack of consolidated control, some positive steps are taken when a particularly committed and connected individual takes charge of individual institutions:

> [We] were lucky that Dr. [X] was then Director General of [one of the transport agencies]. He was a very capable leader and had meetings with the thirteen transport agencies every week (Bangkok).

Institutional cooperation may hinge on the effort and support of a single individual. However, apart from normal departures and retirements, in Global South cities, it is not uncommon for an entire public administration to be replaced with the arrival of a new mayor, governor, or president. Then, all previous plans are overhauled too:

> The national transport plan or public transport plan, for example, changes every time a new government assumes office… We do not really know for sure if any project that we discuss will come to fruition. You have to ask the Prime Minister if it will materialize… Even the director general isn't guaranteed to know because these positions change when a new administration comes into power… This is the Thai style system (Bangkok).

---

[79] For a discussion on how the value capture approach can be used to stimulate TOD and finance rail transit and station area improvements, see Gihring (2009).

> As of now, we only have a National Transport Policy, which covers sustainable urban transport, but it is not yet adopted.[80] Why? Because, unfortunately, the President [Benigno Aquino III] set it aside for the obvious reason that that policy framework was prepared during the term of the previous administration [led by President Gloria Macapagal Arroyo] (Manila).

> Under the former Minister of Transportation,[81] there were regular staff rotations which were announced every other Tuesday and Thursday. It made capacity building difficult. It's hard to keep track of the impact [of our advisory activities] if people are transferred every two weeks. There was no knowledge management involved. That means, when people left a position, they took all of the knowledge with them. Then, we had to start from zero. It was exhausting to present everything from scratch over and over. And more importantly, at the city level, much depends on the term of the mayor. We intentionally choose to work with newly-elected mayors. It would be difficult to work with mayors who are in the middle or at the end of their term as everything may be binned or replaced by something else entirely (Jakarta).

An interviewee contrasts Southeast Asia's situation to Japan's—where there is institutional continuity and long-term planning, which minimizes personal favouritism and vendettas:

> Going back to Japan, they have this fifty-year plan for [transport] infrastructure; they already know where ... the rail lines will be. ... So when the time comes to build the new bullet train line, then people or cities cannot come up with the excuse that they did not know about this (Manila).

Stymying legal framework

While administrators are expected to follow the law, the successful transfer of planning policy approaches from elsewhere requires a degree of openness to new ideas and a willingness to work towards modifying local legal frameworks. As an interviewee puts it:

> Laws are written by people, they can be changed. (Bangkok)

Overall, the Southeast Asian region is mired in bureaucratic rules and regulations, whose intricacy and rigidity thwart policy transfer—and also planning innovation more broadly. The following quote summarizes the problem:

> [Public officials] in Thailand are very weary of legal transgressions. When it comes to new, innovative projects that have never been tried before, governmental agencies are afraid that proceeding with those might get them into trouble with the Office of the Auditor General of Thailand, which is likely to scrutinize anything unfamiliar. Sustainable transport projects tend to fall into this trap. ... They'll say this is illegal, this cannot be done. We start from the barrier. The role of bureaucracy here is to act like a gatekeeper. If they have to take a risk, they will not do it. They'll only follow laws and regulations, but sometimes those very things can obstruct new, innovative policies. Instead of recognizing that there is a public interest objective, then find a way to work toward a legal amendment, what transpires is actually the opposite. They'll flat out refuse... That is so despite the fact that there are [effective] case studies in other cities (Bangkok).

---

[80] The document was finally adopted in 2017.

[81] Referring to Ignasius Jonan, currently Minister for Energy and Mineral Resources.

Among myriad legal barriers in developing mass transit (bus- or rail-based), land expropriation laws and procedures emerge as a major hindrance. Unconducive local laws provide undeserved windfalls for some land owners while barring the public sector from recouping from the land value that it creates:

> Singapore and Hong Kong are excellent models of TOD because they use their mass transport system as an engine of city development. In Japan, we primarily look at Tokyo. Chinese cities – Beijing, Tianjin and Shanghai – are following what Japan has done in this regard. Japan is considered the original model. Yes, Europe has [TOD] too, but it is particularly pronounced in Japan… [In Japan] profits from the sale of [expropriated] land around stations compensate for the construction cost of the mass transit system. The concession company that builds and manages the system captures the added value through land development and advertising in all stations under its jurisdiction. But Thailand cannot emulate that model because our legal framework only allows land expropriation for the construction of the station itself, not for the development of the surrounding areas. The owners of surrounding lots end up benefitting from the development of mass transit, and real estate developers who swoop those lands reap much benefit from the increase in market value. Meanwhile, the public sector borrows money to pay for the mass transit system. That is not sustainable because cost-recovery cannot depend on ticket revenues alone (Bangkok).
>
> Usually land acquisition is a barrier. … Regulations stipulate that before a project starts, land acquisition has to be one hundred percent complete. But in practice this is impossible, so we usually do [construction and land acquisition] in parallel. That's most difficult. Once people know a project is starting, it is harder [and costlier] to free the remaining land (Jakarta).

In Kuala Lumpur and Jakarta, the chief goal in designing new transit lines is not to maximize efficiency and access but rather to avoid obstacles in the land acquisition process:

> Generally, during alignment we try to minimise land acquisition because people get very upset if they have to give up their property (Kuala Lumpur).
>
> The reason why they built an elevated busway is solely because they didn't want to push it politically on land acquisition. They see it as the easier way. And then most of the time the easier way is not the right way; they're just avoiding a problem by creating another problem. By going elevated, the corridor is not really connected to the other corridors. For example, it passes through Block [X], but you cannot transfer. Things like that. Really, really kind of stupid, if we can say it (Jakarta).

Interviewees point out that transport and land-use coordination—a cornerstone of successful TOD—may be easier in Singapore, China, and other authoritarian contexts where the public sector can expropriate land around transit stations without facing much popular resistance and institutional barriers to land acquisition:

> Even though people say that China is becoming increasingly capitalist, socialism still influences much of their land management system. … China has a very clear land-use policy: it does not privilege anyone – with the exception of Beijing, possibly, where the Communist Party members stay. The Chinese look at the big picture first … and then plan accordingly. … Look at Shanghai: the government owns all the land. When big real estate developers negotiate with the government [to build mega projects], they are given a concession… Land speculation is thus very difficult… Usually, [in Bangkok] the government will try to get people to surrender their lands [near stations], and then people will protest for ten years. In the end, there is nothing gained (Bangkok).

But copying the Chinese approach would be impossible in Southeast Asian countries, where private property rights are afforded more protection, market mechanisms are stronger, and forced expropriation (even with compensation) would invite lawsuits. The result is curtailed accessibility:

> [Shanghai's] high-end Science and Technology Museum is linked to a station along the rapid transit system, which allows children to access it. In Bangkok, the National Science Museum is located way outside the city where only few can reach it. While shopping malls are connected to rail stations (Fig. 3.14), universities and other education centres are not located on any transit line where access is easy (Bangkok).

The often-fraught relationships between property owners and the planning sector give the impression that locals are 'anti-rail'. In reality, this attitude is a local manifestation of opposition by locals to acquisition of their land:

> When people protest against a rail project, what they're really against is the construction of a station on their property because they do not want to surrender their land to the government. Their demand is for the station to move away. But they still want rail transport (Bangkok).

Frustration with the current state of affairs is so high that one interviewee in Thailand vents:

Fig. 3.14 Shopping malls and other upscale land uses in Bangkok lobby to have rail stations located in their immediate vicinity and build direct connections to those (e.g., covered, elevated pedestrian bridges, as in the example above)

> I am pro-democracy, but if we want to get things done, it has to be during this military dictatorship (Bangkok).

The view that a democratic system does not yield ideal outcomes in urban planning is shared by many in the region. While it may be shocking to Western readers attached to the idea of public participation and inclusion, in reality post-political and post-democratic practices (dominated by private corporations rather than governments) are beginning to shape urban governance in Western cities as well (MacLeod 2011).

More recently, interviewees report that legal amendments have been proposed in Thailand which would entitle property owners in station catchment zones to benefits from rail projects, rather than subject them to expropriation and eviction. This would pave the way for TOD:

> Our research with the Rockefeller Foundation is examining whether there are ways for communities around rail stations or canal piers to share profit with the government or [transit] operating agencies [by owning stock in transport companies]. If this is possible, then giving up the land around transport nodes may become … more appealing [for owners] (Bangkok).

Legal barriers, or more precisely lacunae in techno-legal competence, are present even when local planners attempt to implement smaller-scale projects, which do not require land expropriation. For example, bikesharing schemes are considered as one of the greatest policy and technology transfer successes; any city that seeks to appear hip, vibrant, and cosmopolitan is compelled to provide at least one bikesharing system (Pojani et al. 2020). Yet, setting up local contracts can be a challenge. One interviewee in Malaysia recounts:

> Malaysian city councils have a very big problem with contracts. They find it hard to implement new projects that have no legal precedent in their cities. Here is an example from our experience doing bikeshare: a city council looked at how to enter into a contract with us and they found no existing text. So what they did was – this is a funny story – they grabbed a 'vending machine' contract, and renamed it 'bikeshare'. Just think about that. The contract that governs vending machines is totally irrelevant for bikeshare schemes. … A contract should be about how a bikeshare project is implemented and run, so planners should be involved in drafting it, not only lawyers. I think this is a trend across our regulatory and planning agencies. It's that all our professionals tend to their own gardens. Planners only look at zoning codes and development approvals (Kuala Lumpur).

The same interviewee offers the following solution:

> Stop working in silos and start forming multidisciplinary committees or multidisciplinary teams. Lawyers have to work with planners, police have to work with planners, engineers have to work with planners (Kuala Lumpur).

Clientelism, corruption, and populism

Clientelism or outright corruption are other deterrents to the successful transfer of sustainable transport concepts. As a widespread practice in Southeast Asia, clientelism depends on relations of social and financial patronage. Corruption and bribery are also deeply embedded, due to 'the low salaries of the civil servants and political leaders, the ample opportunities for corruption in many public agencies, and the low probability of detection and punishment of corrupt offenders' (Quah 2003, p. 1). In the Corruption Perception Index of Transparency International,[82] Thailand, Philippines, Indonesia, and Malaysia, respectively, score 36, 36, 38, and 47 out of 100. (In the Asia Pacific region, this comparison shows 57, 73, 77, and 85 out of 100 for South Korea, Japan, Australia, and Singapore, respectively.) The local population is not surprised nor indignant by corrupt behaviour but rather sees it as 'normal' (Eskenazi 2006). One interviewee explains how clientelism and corruption shifted transport investment from rail to road in Thailand:

> Thailand used to be viewed as a rival to Japan back in the day, and Japan sent their representatives to visit our rail system. But when Thailand entered the era of democracy, because it did not have a good [democratic] foundation, people were easily manipulated by politicians. ... If you notice, politicians in this country were and are [in the construction business]... They only know how to build roads, not rail ... so the plans for the rail system have been overturned in favour of building roads. ... There are major conflicts of interest in the transport sector. ... The projects here are laced with hidden agendas. ... A handful of companies take turns [winning big contracts]. If you are in the field, you'd know that it's orders from the higher ups that determine who gets the concession. It is not difficult at all to fix the bidding process because the method to evaluate proposals is poor. Most bids are just window dressing (Bangkok).

A representative from the Asian Development Bank in the Philippines echoes these observations:

> We have an independent evaluation department, which examines the country assistance program and identifies the lessons learned. ... If you look back at the project completion reports on roads .... the finding is that the governance issue in road infrastructure is very significant, it is very serious. Corruption in procurement and so on. ... [A lesson is] ensuring that projects are contracted out and not subject to collusion, cartels, and corrupt practices (Manila).

Besides construction companies, the automobile industry is another powerful force which lobbies to protect its interests and manipulates both politicians and public opinion:

> Some say that our politicians are bought by car companies (Bangkok).
>
> Limiting personal car use in Jakarta is most challenging... The automotive sector is worried that, if regulations apply, car sales will sink. They argue that this will lead to unemployment. This is the reason the officers in the transport sector have failed to adopt [rules limiting car ownership and use] (Jakarta).

Even where public transport investments prevail over road infrastructure investments, the decision-making is often driven by financial interests rather than a desire to emulate good practices from elsewhere:

> The government is building mass rapid transit at an over-inflated price. That's the way for people to make money. ... The Prime Minister needs to make sure that all his cronies will

---

[82]The index is formulated annually (https://www.transparency.org/cpi2018). It ranks 180 countries by their perceived levels of public sector corruption according to a survey of experts and businesspeople. A scale of 0 to 100 is used, where 0 is 'highly corrupt' and 100 is 'free of corruption'. The figures reported here are for 2018.

> benefit from these mega-projects. … So, I think the government is motivated by this rather than by [international] ideas and policies (Kuala Lumpur).

Populism, which thrives in a region of weak political parties and government institutions (Kenny 2018), presents another set of problems. While not a corrupt practice per se, it is manipulatory in that it encourages politicians to lend support to 'headline-worthy' rather than sustainable urban transport projects in an effort to win wide popularity while pretending to ignore established elite groups:

> Politicians are simply not interested [in transferring sustainable transport concepts from abroad]. They just do what they fancy and what will win them votes; they don't look at facts, they don't look at arguments, they don't look at numbers. It's just a 'flash in the pan' kind of planning. … No policies, just actions. And only populist actions. For example, building highways and making them free to use. People love free stuff! (Kuala Lumpur).

> The National Transportation Policy has been [temporarily] shelved. … It was not at the top agenda of the outgoing administration. Well, to be fair, there were other pressing issues like freedom of information, the reproductive health bill; so of course the NTP was overshadowed. … It was not as glamorous as the other proposed bills. … But now it is becoming more high profile because of the traffic congestion we are experiencing (Manila).

> When [professional planners] go anywhere [on study tour] and come back, [planning] recommendations are submitted to the executive, who acknowledges the report. But what the elected officials do is, they just order this or that (Bangkok).

Urban planning ends up dominated by projects designed at the whim of political leaders keen on gaining or retaining power, whereas long-term, evidence-based plans prepared by professionals are shelved. While politics is always about balancing various interests and policy issues, and satisfying mass constituencies, in the current era of climate emergency, a clear vision that prioritizes sustainable transport among all other public agenda items is needed:

> A sustainable future is not easy to attain because of so many interests, many of them conflicting. The interests of the private car owners, the interests of the car manufacturing sector, the interests of public transport operators, the interests of the government, the interests of public transport riders. But what's important is for the government to be very clear with its policy. What is really the priority? (Manila).

Apathy and inertia

General inertia and a lack of enthusiasm for innovation among those who are in a position to drive policy transfer forward is another oft-cited barrier. An interviewee in Bangkok is rather bitter about this state of affairs:

> The experience of other countries does play a role to a certain extent. When it reaches Thailand, however, people are too lazy to follow through. … Strong bureaucracy has moved [the rest of] Asia forward: look at Japan and Singapore. However, Thai bureaucracy is just a place for the incompetent. … Thai bureaucrats figure that, if they work more, their salary will stay the same. If they work less, their salary will still stay the same. On balance, they might as well work less. … Capable people are pushed out of the public sector… We need motivated government officials, who look at what they do as a legacy for future generations; that's how applicants for public service jobs should be screened. We need those who want to become legends… (Bangkok).

Interviews in Malaysia and the Philippines echo a similar sentiment:

> I think we are not competitive enough, and a lot of people are rather complacent, especially government people, I'm sorry to say that. The government has always been practising the philosophy of 'don't venture into anything new, don't fix it unless it's broken'. That's always been their adage. We always share with them new ideas and they are always a little cautious and will always reply 'Where has this been done before? Are you sure this is going to work?' (Kuala Lumpur).

> There's no leadership in place to make the hard decision to fix the system. Everybody is saying, 'oh no, let's not rock the boat or else they won't vote for us' – things like that (Manila).

This type of critique is hardly unique to Southeast Asia. In Western countries, through the 1960s and 1970s (the height of the rational planning model), many books were written on the lethargic, outdated, and self-righteous state apparatus, which existed merely to serve itself (see, e.g., Benveniste 1977; von Mises 1944). It may be argued that it was, in part, those attacks which led to the dismantling or privatization of many public services, and the embracement of the 'lean and mean' ideology of neo-liberalism. While neo-liberalization reforms are ongoing in Southeast Asia too, and coexist with various forms of state authoritarianism (Nevins and Peluso 2008), planning bureaucracies continue to be slow and apathetic (in some countries more than in others):

> I get to work with counterparts in the Philippines and other ASEAN countries. [In the Philippines] we've been having the same conversation, on the same topic, over and over but it does not get to the next step. We are very good speakers but, when it comes to the action part, there is no follow up. ... ... Vietnamese colleagues, when they hear about something new for the first time, they're just like sponges; they want to learn about it, and then they'll tell you, 'we have funding, just tell us what to do'. And soon the concept can already be implemented. But in the Philippines five years will pass, the concern will stay the same, and if you trace back the PowerPoint presentation that any government official used, the contents will still be the same. Probably, the difference is that in other countries the transport director will say 'we will do this', and that is a commitment. Here it's not like that. The director will say it, but it will not get done (Manila).

Outside capital cities, the situation may be direr due to scarcer human resources, and supervisors' monopoly over networking funds:

> Bureaucrats in the provinces are not interested [in examples from abroad] because their English is not good. But if you have someone to explain these things in Thai, then they are interested. They do not have much opportunity or access because whenever there are study tours, only high-level bureaucrats go [Bangkok].

Other interviewees note that public service employees—in capitals or outside—are not necessarily apathetic by nature but rather are forced into a state of indifference because they need to defer to those with a higher status within an organization:

> Within our [transport department], I'm sad to say, no unit is interested in looking at examples of sustainable transport abroad. They are not interested at all. These people are hand-picked by our chief minister to basically say yes to his policies (Kuala Lumpur).

> My only regret is that the policy-makers from the government side are not well-versed in transport. Normally they are political appointees. We should come up with leaders with a

> solid [professional] background. Most of all, they have to be planners. In Colombia, they don't just elect mayors because they are popular; they pick capable mayors with experience in governance (Manila).

Owing to a lack of interest or expertise in planning among decision-makers, much urban transport development takes place ad hoc, on a project-by-project basis rather than grounded on long-term visions:

> China is a very good example in that it has five-year plans. And all of the provinces really have to follow the five-year plan. Whereas our measures are just stop gaps, not really sustainable. We don't think long term, we don't plan for the future. ... It's all completely piecemeal. Like the Mandaluyong City electric tricycles, that was an ADB-fostered thing.[83] Ayala Center[84] is thinking of building an underground busway or railway between Bonifacio Global City and Makati,[85] simply because you can't get to their development anymore. The national government's PPP activities along EDSA[86] and Taft[87] were autonomous one-offs (Manila).

> I think transport planning is quite haphazard here, to say the least. The government and transport planners have been doing things without following a masterplan for many years. Things are done 'as needed'. Very reactive rather than proactive (Kuala Lumpur).

The task of educating decision-makers and the general public about the current thinking on urban transport sustainability may be seen as appropriate for academics. But due to populistic planning, which is accompanied by a dose of sensationalism, academics tend to shy away from the media:

> The future is about changing the decision-makers' mindset, and academia always has a role in that. ... Unfortunately, many faculty members – myself included – are not so keen on being interviewed by the media because the media seek sensation rather something useful. We worry about being used to create buzz. We'd rather do it quietly, though our papers or through our training programs. In case a person in a decision-making position happens to have taken one of our courses, then we hope that they will push in the direction [of sustainability] (Manila).

It often takes a crisis to shake up organizations and force them to confront issues of transport sustainability:

> When tsunamis and earthquakes happen, these events have more impact in driving policy and governmental action [than abstract knowledge about risks]. This might owe to Thailand's [institutional] culture, which I think is similar to Italian soccer culture. Italian [soccer players] don't plan ahead all that much, but are good at troubleshooting (Bangkok).

There are gentler approaches, however. For example, toolkits, which (a) detail the steps that need to be taken to achieve a particular result, (b) are tailored to the local context, and (c) are created based on local input, can help prod uninterested

[83]Mandaluyong City is one of the municipalities that make up Metro Manila. In 2011, the ADB funded 20 electric tricycles (e-trikes), and has been gradually scaling up that pilot project since.

[84]A major commercial centre in Makati, Metro Manila. It is partly located on Ayala Avenue, a 2 km thoroughfare nicknamed 'the Wall Street of the Philippines'.

[85]Bordering municipalities in Metro Manila. Bonifacio Global City (also known as The Fort) is a major financial and commercial district in the region.

[86]Referring to Epifanio de los Santos, a semi-circular highway which circumscribes Manila.

[87]A major avenue in Metro Manila.

employees into action. The following quote illustrates GIZ's positive experience with creating such a toolkit for use by Indonesian cities:

> In 2014, Bappenas[88] and SUTIP[89] created a toolkit – a sort of technical or practical guide for local government staff – on angkot[90] reforms, parking management, non-motorized transport infrastructure, and emission monitoring (and soon on mass transit). The toolkit is not a boring document – it's fun, with colours and pictures and so on. In our experience, local government staff do not have enough knowledge and skills specific to their job. Usually they have a background in landscape architecture or agronomy and suddenly they have to take care of transport planning. It's better when people want to learn [about sustainable transport, despite their educational background], but mostly they do not. So, with the toolkit, they'll know what to do when their boss gives them directions. They can work more effectively because the toolkit helps them. Not only did we give out the toolkit, but also conducted trainings – though we haven't monitored the post-training situation (Jakarta).

Information bottlenecks

Because of their professional stature, all the participants in this study are well informed about contemporary urban transport concepts. But the population at large may find it difficult to access information through formal government channels—the widespread availability of the Internet notwithstanding:

> You have to send in a formal request for reports or data from [government] agencies. Even with promotional materials on transport policy, it is not easy to just go and pick them up because they are not widely distributed… You always have to make sure you know who your contact is over there so that you could follow up (Bangkok).

Social hierarchies thwart personal and institutional communication, leading to planning deadlocks. The example below is from Indonesia but hierarchical (and patriarchal) relationships are pervasive throughout Southeast Asia, and the region cannot be fathomed without taking those into account:

> The culture in Yogyakarta is such that the Sultan holds the highest authority.[91] So the people or the institutions are afraid of asking him anything, and as a consequence, he doesn't take any action [because he doesn't know what the people want]. … These social-cultural aspects hamper progress in the transportation arena (Jakarta).

Another oft-clogged communication channel is between government organizations and academia:

> There isn't much contact between public agencies and educational institutions. For example, when universities hold a conference, public officials usually do not attend, and the discussion is theoretical. Academics get to meet officials when invited for a talk, or in multi-sector meetings, but these don't happen often. Public forums are an exception; these aren't about knowledge building but more about trying to find mutual ground (Bangkok).

---

[88] Indonesian Ministry of National Development Planning, also called National Development Planning Agency (*Badan Perencanaan Pembangunan Nasional* or Bappenas).

[89] Sustainable Urban Transport Improvement Project (SUTIP), a GIZ project (2009–2016) commissioned by the German Federal Ministry for Economic Cooperation and Development and executed locally by Bappenas.

[90] Minibuses.

[91] Yogyakarta is a special region, ruled by the Yogyakarta Sultanate, the only officially recognized monarchy within the government of Indonesia.

Beyond general inertia, one interviewee offers a more cynical explanation for this lack of communication between researchers and practitioners:

> Academic research is overlooked because it does not produce much kickback. In general, Thai bureaucrats are corrupt and expect to receive kickbacks from projects (Bangkok).

Some interviewees idealize Western countries, claiming that 'research and practice abroad are integrated.' In reality, planning academia and practice often operate in parallel universes, even in countries with a mature planning sector and well-established academic networks (Hurley et al. 2016). In any case, these circumstances are not very conducive to policy awareness or transfer.

Even conferences and seminars, whose very purpose is communication, can leave participants with a feeling that nothing new has been learned which could be transferred. In relatively conformist cultures, this owes, in part, to an imperative to keep the discussions within the 'Overton window' and limit any radical proposals:

> I do not attend seminars in Thailand because communication in those is one-way. We go to listen to the research findings produced by some consultancy that was commissioned by a governmental agency. It will be something that we already know. … Questions and details outside [the window of discourse] are not discussed in depth. I do not want to have a dialogue with a consulting team. There has to be a dialogue with decision-makers, who can set up a strategy and really unlock the obstacles. Or else the people who research will keep researching, the people who plan will keep planning, but nothing will come of it (Bangkok).

The same is observed in some international gatherings:

> Sustainable transport is a global conversation. By chance, when I attended a meeting in Iran, the topic was Iran's transport sustainability. Different cities came to the meeting, … they all envisioned the same activities, projects, and ideas. But I didn't see any city that would tell us how successful they were [in implementing those]. … It was more like they only talked about policy, but not policy in practice (Bangkok).

Ongoing love affair with the car (and the motorcycle)

While the term 'sustainability' has entered the public agenda and discourse, in reality all capitals are still in love with automobility. This is the case at the personal and institutional level. First, let us look at pro-car institutional approaches:

> If the decision-makers have to choose between building a road or a footpath (Fig. 3.15), which one do you think they will choose? … Governmental agencies in Thailand still insist on building new roads. Recently, just last week, the Bangkok Metropolitan Administration (BMA) announced their plan to expropriate more land to build roads. Cars are always prioritized. … (Bangkok)

In Malaysia, automobility has been promoted through explicit government policies which have made car ownership more affordable and have diminished the appeal of public transport:

> In Malaysia we are a little backwards in terms of ideas. It's all about cars. … Malaysians are still stuck in the car-centric age. The driver is usually the king of the road (Fig. 3.16). … And the current car problem owes to our Prime Minister [Mahathir bin Mohamad] who formulated a National Automotive Policy so as to promote the domestic car brand, Proton.

Fig. 3.15 Poorly maintained sidewalks are common. The example above is from Jakarta

> … The National Automotive Policy made it easy for people to buy cars. At one time, the mini buses in Kuala Lumpur were so packed it was difficult to get in. … Once people could afford to buy cars, why would they squeeze themselves into buses? (Kuala Lumpur)

Fig. 3.16 Wide, multi-lane roads dissect Kuala Lumpur

Now Proton is seeking to develop electric and hybrid vehicles, but the transition from conventional cars to these eco models has met with difficulties. It appears that the users are not ready to make the switch:

> When we speak of EVs in this region, the market readiness is not there. The awareness isn't there yet to adopt such an advanced technology. The most frequently asked question by the public is how far can an EV go per charge. They should know that the purpose of an EV is to mitigate the pollution level in our city, not to drive long-distance. We lack charging stations in the Klang Valley because there are virtually no EVs anyway. I know it's a chicken and egg situation. ... The hybrid vehicles that we were developing, we've had to shelve those for a while because of the high price which consumers couldn't afford (Kuala Lumpur).

Where physical interventions are needed, policy transfer often necessitates compromises between spaces dedicated to cars and alternative transport spaces, as city areas are not infinite (Fig. 3.17). Interviewees suggest that active transport usually loses to car-based mobility during those trade-offs:

> Thailand has size specifications for different types of roads. In the past, there was a plan to include a sidewalk lined with trees, etc.... But when we tried to implement this, it was very difficult. Expropriating land from private owners was a challenge... The road space was needed for cars... So we had to sacrifice the sidewalk (Bangkok).

> In Kuala Lumpur, we don't prioritise walking; we build overhead bridges for the pedestrians, which force them to walk up and down stairs. It's a great inconvenience for people who are

Fig. 3.17 Excessive road and parking spaces in Manila

> disabled or old. By contrast, Seoul is quite pedestrian friendly. Their crosswalks are all at-grade – although they have a lot of car traffic too (Kuala Lumpur).

Another interviewee explains how a BRT project has to be modified—thus increasing capital costs—in order not to inconvenience car drivers:

> Because Kuala Lumpur is constrained in terms of road capacity, the authority is not willing to give up space [at grade] for BRT. The willingness isn't there! So we see our system being elevated. Personally, I don't like this because it spoils the aesthetics of the city and does not allow for integration at street level (Kuala Lumpur).

Even where planners and politicians realize that car dependence is not a sustainable practice, automobility is entrenched through decades of privately-driven investment in infrastructure:

> In the Klang Valley, the national government is struggling because they've got to right the wrongs of the past 20 to 30 years. Building highways has only generated more traffic. … The development of highways should be understood in the context of privatisation. Everything is privately driven in this country! Highway projects are merely for profit, they don't solve the problem of traffic congestion. At most, they just transfer traffic congestion further downstream.

At the personal level, years of auto-centric planning have normalized car dependence—certainly among the upper-income strata:

> It is so ingrained in people's minds that roads are just for cars. (Kuala Lumpur)
>
> Once you can afford private vehicles you don't ever think about using jeepneys or any other public transport modes again. Maybe you will take taxis but then it's still a single occupancy vehicle. (Manila)

Even where locals claim to want more public and active transport provision, they are unwilling to sacrifice car ownership and use:

> From what I observe, people in Bangkok dream of becoming Singapore. But they don't know that it costs a lot to own car in Singapore due to government controls… I think Thais want it both ways: easy to drive, easy to walk.

In status-conscious cultures, car ownership is a guaranteed way of demonstrating one's purchasing power and rank in society—and thus be treated accordingly (Ashmore et al. 2019):

> Think about this: if a man wants to impress a woman, having a car is an advantage. (Bangkok)

An interviewee contrasts Manila to Tokyo in this respect:

> My favourite [role model] is Tokyo because their public transportation is a great equalizer.[92] I studied there for four years. Because their public transportation is the fastest and most efficient mode, even presidents of companies will take it. (Manila)

Motorcycle traffic presents a nearly unique problem in Asia, for which there is little precedent in the West:

> Eventually, we have to look at these [Western] models and ask how much can be applied here; these countries do not have so many motorcycles running around like Thailand (Fig. 3.18).

---

[92] According to the World Factbook produced by the Central Intelligence Agency (https://www.cia.gov/library/publications/resources/the-world-factbook/), Japan's Gini Index of income inequality is 37.9. This compares to 36.8 in Indonesia, 44.4 in the Philippines, 44.5 in Thailand, and 46.2 in Malaysia. The Gini Index varies from 0 to 100, where 0 is perfect equality and 100 is perfect inequality.

Fig. 3.18 Heavy motorcycle traffic is typical in the region. The example above is from Jakarta

> … For example, Australian cities have an ok public transport system and I like their tram networks. But in Thailand motorcycles would be a problem if we are to introduce trams as a public transport mode. (Bangkok)

Despite a generally bleak picture, there are sparks of hope:

> What I do is, I conduct education campaigns for children in public schools in Bangkok on traffic discipline (Fig. 3.19), so that they learn which mode of transport to take and how to use mass transport… I don't think that in this generation we will see any significant change, but in next generation, it might be considered cool to pick up your girlfriend and go to Siam Square using public transport instead of a private vehicle (Bangkok).

Sustainable transport is a package

Any specific transport measures cannot yield satisfactory results—or may be entirely unfeasible—if transferred and employed in isolation. It is a well-known fact in the transport field that packages of measures need to be transferred in all cases (Stead and Banister 2003). But, as one interviewee in Thailand says:

> What gets implemented here is only the tip of the iceberg (Bangkok).

Where transfer targets individual policies, it is bound to fail. For example, one cannot limit people's car access without providing any decent mobility alternatives:

Fig. 3.19 Traffic education poster designed by the National Center for Transport Studies at the University of the Philippines, Manila

> In Singapore, cordon pricing limits the number of cars that access the inner city zone. That is something I want to see happening [in Bangkok], but we need the public transport system to make that policy possible. … We cannot just stop people from driving their own cars into the inner city. There is a potential lawsuit waiting for us if we just carelessly implement such a measure (Bangkok).

> You can't just stop people from driving their cars. … If we attempt to do that, even in the City of Manila alone, I'm sure a lot of people will raise hell. Some people will be too selfish about it. The impact will be negative rather than positive. They closed one street in Cebu and the backlash was tremendous. … You have to give people a viable, comfortable option. We haven't got that. There is no way to get people in Manila to leave their cars at home and stand in line for an hour queuing for the train. Or sit in an empty bus for two hours with the driver refusing to move until he fills up the vehicle (Manila).

Mass transit (requiring a massive investment) cannot function well if pedestrian routes that lead to stations (requiring a much smaller investment) are not created or maintained:

> A good example [of inspiration from abroad] are the pedestrian facilities in Singapore. Those are a key to the success of the metro there. We know that every public transport rider is also a pedestrian, users need to walk [to reach stations] as this is not a door to door service. So this was our recommendation for the Jakarta MRT. The PT MRT[93] has to maintain not only their rolling stock but also the areas around stations. It was mandated in the regulations (Jakarta).

[93] A municipal corporation that operates the newly inaugurated MRT system in Jakarta.

Fig. 3.20 Informal settlements along Bangkok waterways

> In European cities, the 'last mile problem'[94] is addressed by providing good walking and cycling facilities [along routes leading to transit stations]. If we want this, we need both policies and infrastructure support. It is not enough to say, 'hey, let's all walk' (Manila).

Once built, a mass transit system needs to be not only maintained but also protected from floods, which are common in the region, and flood protection requires multi-sector cooperation:

> Technically or technologically we really do not have a problem. We are capable of building a subway system. But, we will still experience flooding unless informal settlers are relocated away from estuaries, tributaries, canals, and rivers (Fig. 3.20).[95] So if we have a subway, and there is a flood while you're riding it, there will be a power outage [and you'll be stuck underground] (Manila).

If a policy requires that designated classes are charged lower public transport fares (e.g., riders who are in a more economically vulnerable position), then that policy needs to consider whether such a scheme will lead to discrimination against that class of people—especially in a context where much public transport is informal or very loosely regulated (Fig. 3.21). If so, that policy cannot be adopted unless it is backed

---

[94]Describes a situation where the nearest transit stop is too far to walk and too close to drive (even if finding parking is guaranteed).

[95]Many of Manila's informal settlements are built on waterways. Heavily polluted, these feed into the Manila Bay. Historically, squatters have been neglected, or forced to resettle to distant provinces. More than a decade ago, the Philippines Supreme Court issued a mandate to municipalities in Metro Manila to clean up the waterways, and, as a consequence, the cities started relocating more than 100,000 families (more than half a million inhabitants). In a departure from previous approaches, the aim was to relocate people to nearby units or at least in city. However, the initiative has been controversial; many squatters are resistant to moving, the water quality of the Bay has not improved, and flood risk has not been mitigated.

Fig. 3.21 Informal, polluting tricycles in Jakarta

up by public subsidies. One interviewee in Manila explicates how a fare discount scheme (emulated from more developed economies) can backfire:

> A concern that may not have surfaced when you look at foreign studies is the [discrimination faced by senior citizens]. In the Philippines we require that senior citizens be given a discount on public transport.... But if you notice, no one is really paying for that discount. It is actually the jeepney driver who is to bear the cost of the senior citizen discount. So jeepneys don't pick up senior citizens or they don't give them discounts (Manila).

Clean electric vehicles cannot be deployed if the road conditions are not good enough to support this technology:

> In Thailand, 100 cc [gas] motorcycles are what the market demands...[96] The market for electric motorcycles is very limited... While electric motorcycles cost as much as gas motorcycles, ... their maximum speed is low, and roads in Thailand require high speed and good acceleration. ... It's very different from the situation in China where electric motorcycles are widely used (Bangkok).

Beyond physical infrastructure, sustainable transport initiatives need to be supported through financial mechanisms. For example, fuel must be priced (in other words, taxed) in such a way that discourages individual car use and incentivizes public transport ridership, and car parking cannot be free:

> The government says that they want to support sustainable transport. Meanwhile cars are still allowed anywhere and are entitled to benefits like subsidized fuel and free parking. So how can anything be accomplished? (Bangkok).

---

[96]cc stands for cubic centimetres of displacement; in other words speed; 100 cc is considered average.

> Why is Singapore so much better than anywhere else? Because they charge private vehicles so much, in all sorts of ways. … It's a combination of carrot and stick measures. … And a lot of the money that they collect is used to make sure that one doesn't really need a car, one can use public transport – which is what London is trying to do through their congestion charge (Manila).

In addition, any intervention to improve urban transport needs to go hand in hand with land-use management. Uncontrolled sprawl in the urban fringe and along highways neutralizes the benefits of public transport provision in the core:

> Land use control in Thailand is very difficult and makes any [transport] planning impossible. This is at the root of all problems, and there is no sign that the situation will improve anytime soon. Planning does not lead, but rather follows the development of the city … [and] … the initiatives of the private sector. In the absence of [land use] controls, businesses develop alongside highways: … shopping malls, convention halls, hotels, or condominiums. This leads to a host of problems (Bangkok).

Interviewees recognize that even smaller-scale infrastructure projects, such as bicycle lanes, need to be integrated with land uses if policy transfer is to be successful:

> I think Singapore is up there [as a role model] because of years of extensive and integrated planning. So in Singapore, bicycle lanes are linked in a Park Connector Network [which connects the various parks and other green spaces] (Fig. 3.22) (Kuala Lumpur).

In many cases, a lack of coordination between transport and land-use projects owes to the institutional turf wars discussed earlier. An employee of a transit agency in Bangkok explains how this precludes policy transfer:

> Other agencies do not understand that we are not only responsible for building the rail lines, but also have to attract as many passengers as possible. To do so, we must work in parallel with land use. Urban planners must come talk to us to promote the development of the areas around the stations… There is still much confusion and misunderstanding about this idea. Even now, when I speak about TOD, people say 'why are you meddling in real estate development?' Haven't planners all gone on study tours to Hong Kong to see how TOD is done? They think that Hong Kong [has created TOD] because it has limited land. In reality, only a quarter of Hong Kong's land is built up. Homes are concentrated around the rail stations, and they only invest in infrastructure when strictly necessary. That is clever. In Bangkok, because of urban sprawl we spend much more on infrastructure (Bangkok).

Transport and accessibility are also linked to the issue of informal settlements. While makeshift housing sometimes stands in the way of potential transit lines (Fig. 3.23), squatters are strongly motivated to be within easy reach of work, given their meagre means. Thus, on balance, any squatter evictions would aggravate rather than alleviate a city's transport problems—if one considers all residents, not only the middle classes:

> The conventional practice of past governments has been to move poor people outside of the city to areas developed by the National Housing Authority where the [public] transport system doesn't reach, and there is no way to get into the city. Squatters become worse off than before as a result (Bangkok).

Part of the reason why the cities under study have not been thinking in terms of intervention packages is that urban transport is still dominated by engineers rather than planners:

Fig. 3.22 Segregated bicycle paths in Singapore

> In Thailand, transport is listed under the discipline of engineering, which is our mistake. When the word 'engineering' is attached to transport, most of the participating bodies and actors will come from the field of engineering, with such organizations as the Engineering Institute of Thailand or conferences such as the National Transport Conference (Bangkok).

A 'hard science' bias has hampered sustainability efforts because planners and engineers have competing visions. Planners are more concerned with the overall integration of the built environment, and emphasize the social, ecological, and land-use impacts of engineering works. In contrast, engineers tend to prioritize technical

Fig. 3.23 Kampungs along railway easements in Jakarta

aspects such as traffic flow, high speeds, costs, and efficiency of vehicle movements inside cities:

> In Hiroshima I was given a tour of the cable car project… Here in Thailand, administrators would be against that kind of project. They'd argue that a cable car would obstruct the mountain scenery and would not be cost-efficient because [it's a low-capacity mode]. But the local government in Hiroshima takes the opposite view: they understand that old people who live [up the hills] need the cable car to move around. Although one cable car can only accommodate 20–25 passengers, the project truly responds to people's needs… Why does this not happen in Thailand? Because we lack that kind of attention to the quality of life… we look [at transport] from a cost-efficiency perspective (Bangkok).

As such, most engineers (though not all, obviously) place an excessive emphasis on outcome thus neglecting process:

> [Engineers], they think in terms of 'before and after'. There is no 'in the meantime'. … Increasingly, I have been taking a more political-science approach to what I do, rather than just calculating the project's costs or its engineering feasibility (Bangkok).

> While there is a hard component [to urban transport] there is also a soft component of social change. All these technicians and engineers don't even want to look at that but the soft part is very important (Manila).

Unculturized approaches

To be successful, policy transfer requires sensitivity to local cultures—'culture'

Fig. 3.24 A formal, colonial public space (Merdeka Square, left) contrasts to an indigenous bazaar (Chinatown, right) in Kuala Lumpur

understood in both social and institutional terms. Urban planning varies greatly depending on the place in which it is practiced (Fig. 3.24). Culture shapes the way in which planning systems are devised and constructed, as well as the way in which planning operates and performs. It affects both formal and informal rules, methods, and procedures, and also the way planning interventions are received by local peoples. Differences in culture are reflected in a diversity of professional and social practices (Knieling and Othengrafen 2015), which can hamper policy transfer or lead to the transfer of 'wasteful' policies:

> If something is good, it doesn't necessarily mean that it's right for Thailand. … One of the big problems that we are facing now is that basic infrastructure does not correspond to the [local] context. For example, Thai people are very sociable (Fig. 3.25), but when you build something that prevents people from going to see each other, that in itself is against [social] sustainability (Bangkok).

The following quote points to the fact that planning cultures and traditions are forged over decades or possibly centuries, and it is difficult for cities which have been born out of informality to embrace 'rational' Western-style planning:

> Historically [since the days of British settlement], Australian cities have had proper planning. We don't have that kind of tradition. Kuala Lumpur started as a hamlet at the muddy confluence of Klang River and Gombak River, where all the boats would gather. (Kuala Lumpur means 'muddy confluence' in Malay.) From that spot, at the Masjid Jamek Mosque the coolies and immigrants started to build up this city up without proper planning (Kuala Lumpur).

Fig. 3.25 Sidewalks as social space in Bangkok

But even transferring discrete elements—sometimes as basic as a roundabout or a bus ticketing machine—rather than a whole planning system requires consideration of local practices and behaviours:

> We received funding from the European Commission … [to improve] road safety standards, particularly with respect to traffic lights and roundabouts. … We really questioned roundabouts' probability of success in Thailand because being Thai, we know the behaviour of drivers here. If there is no traffic light, then they won't stop. [The European partners] wanted to convince us. They kept insisting that it was possible and took us to see a case study (Bangkok).

> We all know that Japan has a 99.9% punctuality rate in public transport. Even if we import their technology, we can't achieve a 99.9% punctuality rate. You have to know that in Malaysia – all over Southeast Asia – this is quite impossible because of our culture. In Japan, people pay their fare before they get on the bus, so the bus stops for just one minute. You have two doors on the bus, one to enter and one to exit, so you save time. Japanese society is very honest and trusting. Here in Malaysia it's different. You have to queue and pay on the bus, extending the stopping time to four or five minutes, because if you allow people to pay before they board, then everyone will evade the fare. If you have two doors on the bus, people will get in from the back door and won't pay. Therefore, our buses aren't punctual (Kuala Lumpur).

> Women in the Philippines do a lot of shopping so there should be an area for luggage on public transport vehicles (Fig. 3.26) (Manila).

Fig. 3.26 Women-only area on TransJakarta

Other interviewees explain Bangkok's and Manila's 'undisciplined' traffic behaviour in terms of local cultural practices:

> We lack strict adherence to traffic laws… because of the local culture… [and] social values of Thai people… For example, when you drive, you do not drive in your designated lane, you park in a prohibited area (Fig. 3.27), you invade the BRT or public bus lane (Fig. 3.28), among others… Legally speaking, [motorcycles] have to stay on the left lane, but in reality they go everywhere (Fig. 3.29). … Vendors occupy bike lanes and use them for their stalls… It's a problem of lack of discipline. The saying that 'doing whatever you want is real Thai' is applicable here. This is a problem in all sectors, not just transport (Bangkok).
>
> I admire Singapore, although they call it the City of Fines. It would be an advantage if we also [imposed more fines for traffic misconduct], because it would improve people's behaviour and, at the same time, increase the local government revenues. … But knowing the cultural and political behaviour in this country, it will be difficult to emulate Singapore. … My understanding of Filipinos is that we don't like to be disciplined. Freedom first. … We're so used to getting on and off the jeepneys anywhere, we don't follow jeepney stops (Manila).
>
> In Kuala Lumpur, [law] enforcement is quite good but generally Malaysians have poor road ethics. We double park everywhere, drive in the emergency lane, don't really care about [Council] rules. Something is wrong with our education system (Kuala Lumpur).

As expected, these practices lead to traffic congestion, delays, and unpredictability. In Kuala Lumpur, an interviewee ascribes locals' lack of concern about these issues to a different cultural understanding of the concept of 'time' than in the West:

Fig. 3.27 Cars parked on the sidewalk in Jakarta

> To people overseas, time is money. Here, people are more laid-back and do not value time as much, because wages are low.

In time-rich and cash-poor cities in Southeast Asia, punctuality may indeed be irrelevant. But studies have revealed that the time wasted in traffic is a major annoyance, especially for middle- and higher-income individuals (see Wu and Pojani 2016). Hence efforts to promote alternative transport should capitalize on the liberation from driving stress that collective or active modes can provide.

From an institutional perspective, a cultural aspect that hinders policy transfer efforts in the Southeast Asian region is politicians' and bureaucrats' preference for minimizing public criticism and confrontation. This explains, for example, the limited success in implementing a BRT system in Bangkok:

> What the government wants is not maximum efficiency. Instead, what they want is a minimum level of public criticism… This is one of the principles of policy formulation in Thailand, and BRT [in Bangkok] is one such example. Prior to BRT implementation, a team of experts proposed a different route, but that proposal was discarded. The reason given was that, although the proposed route would be more efficient, it would run into more popular resistance (Bangkok).

In multi-ethnic countries, such as Malaysia, tensions between population sub-groups hamper sustainable transport interventions in unexpected ways:

Fig. 3.28 Cars invading the BRT lane in Bangkok

> We’ve trialled a pedestrianisation project, but the local merchants were not too happy because they believe car traffic brings in customers. Malaysians like staying in a cool environment and they don’t like walking under the hot sun. We have tried many times to engage with merchants and local communities but discussions in Malaysia are always racially tinged. So the issue turns into a ‘Malay government vs. Chinese businessmen’ confrontation, which in reality it is not. … If only we could put aside racial biases and appreciate our own diversity… (Kuala Lumpur).

Where international teams are involved in the same project, different working styles (which also depend on culture) may preclude successful adoption or implementation. While outsiders tend to lump all East and Southeast Asian cultures together, cultural differences exist and make collaboration difficult for professionals from different countries in the region. While stereotypes about Japanese restraint abound in the West, one Thai interviewee offers the following insight:

> I think that the Japanese [working style] is straightforward, unlike how Thais do things, which usually is not straightforward. … [Thai] bureaucrats, in many instances, are afraid to speak up, especially if their boss comes along. I have seen twenty local bureaucrats sitting silent the whole time because of the presence of their boss. This is very typical of Thai bureaucracy (Bangkok).

Personal connections are key in Southeast Asian professional settings:

> The [Southeast] Asian system is based more on social networks, rather than institutional relationships like in Europe. … There is a need to meet and talk to each other in person.

Fig. 3.29 Motorcycles encroaching upon bicycle lanes in Manila

> It involves [personal] relationships, frequent visits, and friendship. … It's about going to seminars together, as well as sharing challenges and opportunities, in order to create a push for sustainability. … It's not *what* you know, but *who* you know (Bangkok, emphasis added).

It must be noted that some characteristics of Southeast Asian cities, which outsiders may ascribe to 'culture', are perhaps due to other factors such as the local economy. For example, street vending (Fig. 3.30)—a characteristic of all cities under study—interferes with pedestrian and cycling transport. Vendors end up obstructing new cycling lanes or sidewalks (see Panjaitan et al. 2020). But this portion of the informal economy may disappear once cities strengthen their economic base and transform their labour practices (i.e., shorten office working hours, adopt telecommuting, and the like):

> We see street food vendors as something [culturally] unique to Thailand and other [Southeast Asian] countries, but the underlying reason is the low wages and a lack of free time to go grocery shopping and prepare food [at home]. So office workers have to buy food on their way to work. … I have been to Zimbabwe when their economy collapsed, and there were many street vendors, like in Bangkok. … If Silicon Valley, for instance, drives down the wages and takes away road space, it might become like Bangkok, who knows? (Bangkok).

Similarly, an apparent lack of concern with transport-related health and safety—much emphasized in more developed settings—may not be due to 'culture' but simply to poverty:

Fig. 3.30 Street food vendors provide a valuable service for residents but also obstruct sidewalks. The example above is from Kuala Lumpur

> In Marikina, we do require that bicycle riders wear helmets. But the problem here is the economic condition of the community. If you issue a fine of PhP500 [$10] for not wearing a helmet to someone who is a poor worker, he will say that the reason why he rides a bike is to lower his transportation costs; where will he get the money to pay for both the fine and the helmet? (Manila)

Partly due to a lack of cultural understanding, some of the technical advice bestowed by international organizations are too broad and vague. As such, it can apply to nearly any place but local staff find it difficult and time-consuming to operationalize it to the urban or site level:

> We have worked closely with the Asian Development Bank and received technical assistance, or studies, from them. But their writing is usually embedded within theoretical frameworks that are too general for Thailand and thus cannot really be effective in implementation. For example, the report will say things like 'the rail authority cannot be both a regulator and an operator' – but people already know that. But when it's time to implement and you have to negotiate with the union, how do you handle that? Besides, we are overloaded and don't have time to follow through with these reports. They are like the yellow pages: no one reads them (Bangkok).

International donors and policy advisors in the region are becoming aware that such 'policy gifts' (Pojani and Stead 2018b) are unlikely to be implemented:

> It is all very well that JICA is financing another study. But it is not owned by the government so it is not going to be implemented (Manila).

> Broadly speaking, if the government does not ask for [a policy recommendation or intervention] in the first place, no matter how hard we work to push it forward, it will remain on paper. A good advisory activity requires a 'need assessment' from the outset. Does the government need it? If not, then there's no point doing it. As a donor organization, we can't just come to them and force our approach down their throats. They wouldn't accept it. ... For example, GIZ provided assistance in compiling reports on emissions for a number of cities. But we didn't know whether Indonesia wanted to publish such data. As it turned out, the data were never made available to the public because the municipalities did not want that (Jakarta).

Local interviewees indicate a preference for detailed guidance on specific issues, Southeast Asian cities have themselves identified instead of high-level reports:

> Bang Sue Central Station [in Bangkok] is designated to become a railway hub. We asked JICA to help with the management of operations and traffic within the station, so that Bang Sue Central Station can become like the Tokyo Central Railway Station [Chiyoda]. That's the kind of technical assistance we prefer, as opposed to some general, one-size-fits-all plan that we don't need (Bangkok).

Hot, supersize cities

East and Southeast Asia are home to most of the world's megacities. These are relatively new urban systems, for which the West has very little precedent. Interviewees recognize that the differences in sheer urban sizes—and also in land-use mix—preclude policy transfer from Western settings:

> What works in Germany or Holland or Belgium, can that really work in Asia? I don't think so, to be honest. So you need people who have worked here or are from here to understand what a city of 12, 15, 20 million people is like. It is very different from a city of a few hundred thousand (Manila).

Even cities that are motivated to emulate the West (e.g., due to colonial legacies) are unable to do so due to difference in physical size. An interviewee in Malaysia (a former British colony) offers:

> Our rules and regulations were transferred from the British when we gained independence. Even now, we still look at the Anglosphere and [other former British colonies] for inspiration in terms planning policies. However, when it comes to implementation, we don't follow exactly what Westerners are doing because ... we have large, crowded cities with very mixed land uses; that's the traditional way of life in Asia. We have a different approach to transportation (Kuala Lumpur).

In this sense, other Asian examples are viewed as more comparable:

> Guangzhou is a big city with more than 10 million people. If we visit its BRT system, and compare it Jakarta's, it's an apples-to-apples comparison. Also their socioeconomic conditions are more relatable (Jakarta).

While Southeast Asian capitals are much larger than the 'typical' Western city, they have adapted, over the course of decades, to small and flexible public transport modes (Mateo-Babiano 2016). This contrast between city size and public transport vehicle capacity may be incomprehensible to visitors used to conceptualizing megacity public transport systems as heavy-duty subways, complemented by LRT and BRT. Yet, one interviewee in Manila observes that the resilience of indigenous micro transport demonstrates its suitability for Southeast Asian megacities (as a feeder mode, at least):

> What is the most sustainable transport mode in the Philippines? It is the jeepney. The Philippine jeepney has been around for the past fifty years without having been dramatically changed. It presents a lot of problems in terms of safety, efficiency, comfort, etc. but, despite all this, it has survived and is still the main transport mode here. It has withstood all administrations. So why is the government trying to phase it out? Can't they just modernize the engines? (Manila)

Beyond public transport networks, the issue of urban scale is relevant for smaller planning interventions at the local level too. An interviewee in Bangkok, who is interested in the Dutch 'naked streets' concept,[97] says:

> This is a case where context matters because the Netherlands is much smaller than Thailand; its total population is lower than in the Bangkok metropolitan region. Dutch cities do not have much car traffic. That is why applying their ['naked streets'] concept in our context is very difficult (Bangkok).

In addition to size, some interviewees cite their city's hot and humid climate as a deterrent to the transfer of active transport (walking and cycling) policies and infrastructure from temperate climates:

> The one big factor that differentiates Japan and China from us is, of course, the weather (Manila).

There is concern that locals—especially the middle and upper classes accustomed to air conditioning—may complain if they have to exert themselves under the sun. However, recent research on cycling in Singapore suggests that 'to people acclimatised to the tropics, the local climate and the weather variations therein pose less of a barrier than factors such as safety and convenience, which are nearly universal. This is because they have a higher tolerance to heat than residents of temperate climates. It appears that, to make cycling irresistible in tropical climates, similar setups and approaches proposed or implemented in temperate climates are needed' (Lee and Pojani 2019, p. 1).

Some local interviewees support this finding:

> When we talk to people about bikeshare, no one believes it will work in Malaysia because of the temperature or climate. … But the climate hasn't changed much since the 60's and 70's when people used to cycle a lot. Yes, we have global warming but the temperature hasn't changed so drastically. What's changed is that our roads are now crowded with cars… all these engines sitting around, generating a lot of heat. … We have to correct people's mindset that walking and cycling are actually not that dangerous, and our climate is not that hot. Next door, Singapore is promoting walking and cycling (Kuala Lumpur).
>
> In Tokyo, because there are no tricycles, you are forced to walk or bike. I hear people say here that it is too hot to walk or bike in Manila, but summer in Japan is also hot and humid and people walk and bike anyways. That is the most efficient way, plus they have bike paths and bike parking racks (Manila).

[97] The concept involves removing traffic lights along minor roads to improve flow and safety.

Other interviewees suggest that walking and cycling can become much more viable and attractive as transport (rather than merely recreation) in tropical megacities with some thoughtful but relatively minor adjustments to the standard infrastructure design employed in temperate climates (Fig. 3.31):

> We need to provide more shaded facilities to facilitate cycling and walking under the hot weather. We need to design more pocket parks for cyclists and pedestrians to rest, or even toilets and showers [along cycling lanes and walking paths]. Pocket parks can be potentially used for loitering, so to safeguard cyclists they must be planned carefully in terms of visibility (Kuala Lumpur).

Fig. 3.31 Covered walkways in Singapore

3.6 Degree of Policy Transfer

> To what extent are policies transferred? Is transfer soft or hard, and are there instances of transfer failure?

Overall, exposure to, and consideration of, international examples, inspires locals a great deal. Interviewees feel that exposure to a new place or a novel way of approaching particular urban transport issues has affected their mindset. The following quote is typical:

> In urban planning we have to be open to new developments. The more I am exposed to diverse fields and ideas, the more I can connect various pieces together and understand the bigger picture (Bangkok).

Innovations involving exclusively software are more easily transferred than anything that requires physical construction:

> I had a discussion with an app maker [in Bangkok]; he is part of a network of programmers in Singapore. ... He created an algorithm that identifies local streets and alleyways for bike commuting instead [of the main roads suggested by Google Maps]. This application allows one to ride around the whole city (Fig. 3.32) (Bangkok).

But it is rare that a planning policy or tool, or a piece of infrastructure is transferred in its entirety. Only minor interventions with limited impact—such as promotional events—are copied in full:

> Around ten years ago, we initiated the Car Free Day in response to the global trend at the time... But we soon recognized that the Car Free Day alone would not be enough, and that there had to be something more to achieve sustainability and change people's behaviour, not just a one-day event (Bangkok).

Patchy implementation, partial transfer, 'blind spots', and lack of maintenance beyond 'ribbon cutting' are typical throughout the case studies:

> Tokyo and Singapore have excellent rail systems, which allow their residents to forget about commuting by car. ... We have been working continually [to meet that benchmark], but the way that we work does not respond to people's needs, whether in terms of routing, budget, etc. Sometimes, we get the [right] route, but do not have sufficient budget, for example (Bangkok).
>
> For bikeways, I think Marikina City[98] is a good example ... Of course it is too far out if we are to compare it with Copenhagen or Amsterdam. ... Also, the project needs to be sustained – I think some of those bikeways are already being used for other purposes (Manila).

In some cases, locals have sought to combine incompatible tools or policies borrowed from different places:

---

[98] A municipality in Metro Manila. The city built 52 km of bikeways in the mid-2000s, funded by a World Bank grant (US$1.3 million).

Fig. 3.32 Narrow alleys, which, if upgraded, can be used as pedestrian and cyclist connectors, coexist with large collector roads in all four cities. The example above is from Jakarta

> A case in point is the SCTEx (Subic-Clark-Tarlac Expressway).[99] This is subdivided into two segments. One was done by Japan and the other by Australia. ... If you look at the traffic signs, there's a difference in standards. I think there has to be a coordinating agency which harmonizes signs; otherwise it is very confusing for the users (Manila).

> Think about this: the rail system here uses an American safety standard[100] [which discourages development around stations], but they try to design the system to be like Hong Kong's or Japan's. ... If you want your transport system to be an American-style system, then go for it, but remember that in America, rail systems rely on state subsidies to cover their operating costs. If you want to be like Hong Kong or Japan, whose urban rail systems actually yield profit, then you have to follow their safety standard (Bangkok).

Purposely selective transfer is quite common, as policy makers cherry pick those elements that suit their agenda and use them to frame policy discourses (Bulkeley 2006). For example, one interviewee explains why Bangkok has chosen to invest in costly and shiny pedestrian skywalks in the manner of Singapore rather than fixing sidewalks at grade level:

> That's why [the City] is building skywalks left and right: because they look good, and looking good can well translate into construction [budget]. ... Some projects are the result of the foreign private sector deeming them worthy of their investment and thus proposing them to the [Thai] government (Bangkok).

Unfortunately, any 'fixes' or add-ons in case of partial implementation are never as satisfactory or as feasible as they would have been if they were included in the original plan:

> Mass transit and rail projects are all examples of policy transfer, no matter how incomplete. ... When the first mass transit system was built in Bangkok, the question of access, i.e., how people would travel from their home to the nearest station, did not receive much attention. Later, this became a problem that planners had to fix. It turned out that, because the transit fare was high, the primary group of riders were those on higher incomes, who usually owned cars and drove to the stations. So the Office of Transport and Traffic Policy and Planning (OTP) had to find space to create park-and-ride lots... (Bangkok).

Transfer failure is evident too:

> When we build bike lanes, we only build hardscape; we do not plant trees, which makes biking really difficult due to the heat... In [other countries], bike lanes come with greenery as well as facilities like showers, bike racks. And they are connected to other transport modes that accommodate bikes (Fig. 3.33) (Bangkok).

> The law penalizing private vehicles that emit pollution: in other countries, there is usually a database that supports the effective implementation of the law, but in our case, when we locate cars emitting black smoke, there is no mechanism in place to record those and stop car owners from actually continuing to emit black smoke (Bangkok).

And at times, a borrowed policy is adopted only because of personal connections between transfer actors rather than formal, institutional decisions:

---

[99] The longest expressway in the Philippines, opened in 2008. It starts and ends north of Manila.

[100] The National Fire Protection Association standard. NFPA 130 is the Standard for Fixed Guideway Transit and Passenger Rail Systems.

Fig. 3.33 Unprotected and disconnected bicycle lanes in Bangkok's historical centre

> We prepared a report on Intelligent Transport Systems (ITS), which are suitable for the Toll Road Regulatory Agency.[101] Back then, the agency director said he'd implement our recommendations. It's maybe because I have a personal connection to the director. He was my former boss, so I could talk to him directly. But before that time, I don't know whether there was any follow-up to our reports (Jakarta).

Interviewees point out that, even where a policy, program or tool is eventually transferred (more or less), a large time lag may be expected:

> You want to gauge [implementation times] based on the experiences of other countries. … For example how long was the gestation period for a similar policy [or project] in a developed country. If it was five years, don't expect a developing country to do it in the same amount of time – unless you provide intensive support, capacity, financing, technology, and all that (Manila).

While implementation is patchy, virtually all interviewees are adamant: foreign models need to be deconstructed into simpler components and adapted to the local context; they can never be 'copy-pasted' wholesale. For anything to be 'adopted' it must be 'adapted' first. Most organizations understand this, and the following quote is typical:

> [In the past] we used to ponder how we could transfer a policy or plan from abroad in a manner consistent with the original case. But now we are more interested in the components of best practice examples. Which aspect of those can we make use of? … Just copying

[101] In Indonesian: Badan Pengatur Jalan Tol (BPJT).

> and pasting is not enough. Without a body of knowledge derived from a careful learning process and suitable adjustment to the local context, policy outputs will be just a scratch on the surface. … Even though we may look at a certain [foreign] city for inspiration, we still have to adjust their model to the city that we are planning… During the visioning stage, we produce a report comparing select cities, some of them visited during study tours. But when we get to the implementation stage, we have to come up with our own original plan. … More importantly, we now have to really figure out what the mechanisms are that we can deploy to push for certain policy initiatives. For example, if [local] laws are a barrier, how do we amend them? (Bangkok)

Moreover, policy transfer actors have become much more judicious about examples from abroad. They are no longer in awe of cities in more advanced economies:

> Fukuoka [Japan] is one of the most liveable cities. But it is not like I want Bangkok to have everything that Fukuoka has. What I do is, I think about what is good about Fukuoka; why is that so; how have they managed to get to this point, overcome barriers… (Bangkok).

> We don't want to be like Beijing or Shanghai which have grown so clinical that they have already lost their soul (Manila).

Reported earlier were comments about the desirability of Hong Kong- or Tokyo-style TOD in the region. However, the interviewees are also aware that these cities' approach has shortcomings too, and cannot be copied wholesale. For example, an interviewee in Kuala Lumpur underscores that Hong Kong's hyper-dense TODs are inappropriate in a lower-density, tropical city:

> Of course, we cannot replicate every aspect of Hong Kong's TODs because those are too dense and have too much concrete. High-rise buildings are full of air-conditioning units and during the summer, the whole city is like an oven. Our focus now is on linking our mass rapid transit stations to residential neighbourhoods so that riding is convenient for the residents, and parking spaces can also be reduced (Kuala Lumpur).

Complex concepts such as BRT require major adaptations at the implementation stage (and beyond) in order to fit the context of a new city. A representative from ITDP, the organization that assisted with the creation of the TransJakarta system, details the complexities involved in a single BRT element (stations):

> What we suggest requires technical and financial adjustments. And on top, there are the political issues of course. For example, our TransJakarta station designs cost, at the time (in 2013) about $600,000 each. That's a lot of money, way more than Jakarta had ever spent on a bus station. And so, the governor said that the Jakarta city government could not afford that expense, but they had an idea: 'Why don't we look for funding from private entities? Maybe they can run advertisements at the stations'. Regarding the technical aspects, we've been working with the Jakarta transportation agency, and they haven't understood the station design concept well. We need overtaking lanes at the stations [to avoid bus bunching and to allow express services to skip smaller stations]. And that's a major technical issue – it's not a political thing. The transportation agency has to understand it very well in order to be able push for another BRT lane at the stations. … Moreover, the budgeting scheme set up by the government is not really flexible: for example, if a station window is broken, they budget the repair cost for the following year. It cannot be that way (Jakarta).

Adjustment and adaptation are necessary even in the case of seemingly uniform transport products such as rail stations:

> Waiting space [at rail stations] is what I am interested in. At Hua Lampong Railway Station[102] there are not many trains coming in and out. People have to wait; they do not know when their train will arrive. Recently, I conducted an experiment: with a group of university students, we went to the station, moved the benches in the main waiting area to the side, laid down mats and bean bags, and just hung out in that hall. Many people came to sit with us. One German tourist told us that he really liked the idea... He said that in Germany, because the trains are on time, a big, social waiting space is not needed (Bangkok).

An interviewee woefully reflects on the crucial role of context in determining the success (or failure) of policy transfer:

> When we look at the success of others, we neglect to question whether that success can be replicated here in Thailand. Oftentimes we have selected policies to be transferred without careful consideration, which is a double-edged sword. For example, BRT is a project that we did not consider carefully enough to understand the factors contributing to its success in Bogotá, Seoul, and other cities. We only knew that BRT is a 'success story', but not the whys and hows of that success. We took on the project without sufficient knowledge of the concept, of how prepared we were, of how different was our problem, environment, behaviour of transport users... Thai people's behaviour is unlike others, and we have to live with it... Because we rushed into [BRT], the project failed, and we lost a lot of our investment. (Bangkok)

Expanding further on this point, the interviewee below laments the paucity of best practice guides, manuals, and compilations prepared by international aid organizations, which are often striped of context and therefore rather useless to experienced and informed practitioners:

> [Best practice manuals] contain what we already know. No guidebook tells you what Mr. Peñalosa [the Mayor of Bogotá] did to make transport innovations happen. Missing is the discussion of the real costs that would allow decision-makers to take a risk. What gets published is usually an icing on the cake that stresses the importance of sustainability and shows examples like cordon pricing in London. But they don't tell you that it took forty years to accomplish such a feat in London and what the key factors are that prevent or enable such a result. ... And the case of bus reforms in South Korea: everybody just reads about the need for reform and then, bam! The reform. Nobody wants to report the struggling involved and how difficult it was to achieve such a paradigm shift. ... Usually we just see a snapshot of something that was really horrible back then, but now it is fixed! By some miracle or something... It is not canned food, and there is no recipe for success (Bangkok).

Existing studies also cast doubt on the value of exchanging 'best practices'. Rarely do efforts to identify, disseminate, and promote international or national flagship examples contribute to intra- and cross-national learning, lead to improvements in policy and practice, or help avoid newcomer costs (Marsden and Stead 2011; Stead 2012). In-depth case study reports, which discuss implementation barriers and detail pre-tested approaches to overcome those, may be much more useful to local practitioners (Flyvbjerg 2006). However, one wonders whether a focus on contextual differences—'context, context, context'—is merely a facile excuse for inaction on part of potential policy borrowers.

[102]This is the main railway station in Bangkok. Housed in a historic building, it is in the city centre (Pathum Wan District).

References

Ashmore D, Pojani D, Thoreau R, Christie N, Tyler N (2019) Gauging differences in public transport symbolism across national cultures: implications for policy development and transfer. J Transp Geogr 77:26–38

Benveniste G (1977) Bureaucracy. Boyd & Fraser, San Francisco, CA

Bignon E, Pojani D (2018) River-based public transport: why won't Paris jump on board? Case Stud Transp Policy 6(2):200–205

Bulkeley H (2006) Urban sustainability: learning from best practice?". Environ Plann A 38:1029–1044

Chayacani Y, Pojani D (2017) Barriers to successful bus rapid transit expansion: developed cities versus developing megacities. Case Stud Transp Policy 5(2):254–266

Circella G, Tiedeman K, Handy S, Alemi F, Mokhtarian P (2016) What affects Millennials' mobility? Part I: investigating the environmental concerns, lifestyles, mobility-related attitudes and adoption of technology of young adults in California. Working paper, UC Davis, National Center for Sustainable Transportation. https://escholarship.org/uc/item/6wm51523. Accessed 26 Oct 19

Cochrane A (2012) Researching the geographies of policy mobility: confronting the methodological challenges. Environ Plann A 44:5–12

Cook I (2008) Mobilising urban policies: the policy transfer of US business improvement districts to England and Wales". Urban Studies 45(4):773–795

European Commission (1999) European spatial development perspective: towards balanced and sustainable development of the territory of the European Union. Report, Office for Official Publications of the European Communities, Luxembourg. https://ec.europa.eu/regional_policy/sources/docoffic/official/reports/pdf/sum_en.pdf. Accessed 27 Oct 19

European Commission (2002) Asia Pro Eco, Environmental Co-operation Programme: a new initiative for Asia launched by EuropeAid. Press release. https://europa.eu/rapid/press-release_IP-02-1688_en.htm. Accessed 4 Sep 19

Eskenazi J (2006) "Automobile dependence in Bangkok: Trying to change the habits of Bangkok's drivers." Report, German Cooperation Agency (GIZ), Bangkok, Thailand

Evans M (2009a) Policy transfer in critical perspective. Policy Stud 30(3):243–268

Evans M (2009b) New directions in the study of policy transfer. Policy Stud 30(3):237–241

Flyvbjerg B (2006) Five misunderstandings about case-study research. Qual Inquiry 12(2):219–245

Friedmann J (1986) The world city hypothesis. Dev Change 17(1):69–83

Gihring T (2009) The value capture approach to stimulating Transit Oriented Development and financing transit station area improvements. Report, Victoria Transport Policy Institute. https://www.vtpi.org/gihing_tod.pdf. Accessed 28 Oct 19

Gilbert A (2008) Bus rapid transit: is transmilenio a miracle cure? Transp Rev 28(4):439–467

Gómez Rodríguez LD, Herrera Puyana ÁM, García Fonseca MF (2017) ¿Por qué TransMilenio en Bogotá está en crisis? (Why is TransMilenio in Bogotá in a crisis?) Revista Ciudades, Estados y Política 4(3):103–118

Hurley J, Lamker CW, Taylor EJ, Stead D, Hellmich M, Lange L, Rowe H, Beeck S, Phibbs P, Forsyth A (2016) Exchange between researchers and practitioners in urban planning: achievable objective or a bridge too far?/The use of academic research in planning practice: who, what, where, when and how?/Bridging research and practice through collaboration: lessons from a joint working group/Getting the relationship between researchers and practitioners working/Art and urban planning: stimulating researcher, practitioner and community engagement/Collaboration between researchers and practitioners: Political and bureaucratic issues/Investigating Research/Conclusion: Breaking down barriers through international practice? Plann Theory Pract 17(3):447–473

Isaacs R (2000) The urban picturesque: an aesthetic experience of urban pedestrian places. J Urban Design 5(2):145–180

Kenny P (2018) Populism in Southeast Asia. Cambridge University Press, Cambridge, UK

Khalaj F, Pojani D, Sipe N, Corcoran J (2020) Why are cities removing their freeways? A systematic review of the literature. Transp Rev
Knieling J, Othengrafen F (2015) Planning culture: a concept to explain the evolution of planning policies and processes in Europe? Eur Plan Stud 23(11):2133–2147
Krashinsky M (1997) Stakeholder theories of the non-profit sector: one cut at the economic literature. Voluntas 8(2):149–161
Lee Y-S, Hwang E-J (2012) Global urban frontiers through policy transfer? Unpacking Seoul's creative city programmes. Urban Stud 49(13):2817–2837
Lee QY, Pojani D (2019) Making cycling irresistible in tropical climates?. Policy Design and Practice, Views from Singapore. https://doi.org/10.1080/25741292.2019.1665857
Lin J-J, Yang S-H (2019) Proximity to metro stations and commercial gentrification. Transp Policy 77:79–89
MacLeod G (2011) Urban politics reconsidered: growth machine to post-democratic city? Urban Stud 48(12):2629–2660
Marsden G, Stead D (2011) Policy transfer and learning in the field of transport: a review of concepts and evidence. Transp Policy 18:492–500
Mateo-Babiano I (2016) Indigeneity of transport in developing cities. Inter Plann Stud 21(2):132–147
McCann E (2013) Policy boosterism, policy mobilities, and the extrospective city. Urban Geogr 34(1):5–29
Nevins J, Peluso NL (2008) Taking Southeast Asia to market: commodities, nature, and people in the neoliberal age, 2nd edn. Cornell University Press, Ithaca, NY
Nguyen MH, Pojani D (2018) Why do some BRT systems in the Global South fail to perform or expand? In Y. Shiftan, M. Kamargianni (eds) Preparing for the new era of transport policies: learning from experience, Ch. 2. Elsevier, Amsterdam, pp 35–62
Panjaitan TW, Pojani D, Darchen S (2020) Global homogenization of public space? a comparison of western and eastern contexts. In: Mehta V, Palazzo D (eds) The Routledge companion to public space. Routledge, New York and London
Pojani D (2014) Bus Rapid Transit (BRT) (planning and policy). In: Garrett M (ed) Encyclopedia of transportation: social science and policy. Sage, Thousand Oaks, Ca
Pojani D, Stead D (2015a) Going Dutch? The export of sustainable land-use and transport planning concepts from the Netherlands. Urban Stud 52(9):1558–1576
Pojani D, Stead D (2015b) Transit-oriented design in the Netherlands. J Plann Educ Res 35(2):131–144
Pojani D, Stead D (2015c) Sustainable urban transport in the developing world: beyond megacities. Sustainability 7:7784–7805
Pojani D, Stead D (2018a) Past, present and future of Transit-Oriented Development in three European capital city regions. In Y. Shiftan, M. Kamargianni (eds) Preparing for the new era of transport policies: learning from experience, Ch. 4. Elsevier, London, pp 93–118
Pojani D, Stead D (2018b) When West-East planning policy transfer fails to gain traction. J Environ Plann Manage 62(8):1402–1419
Pojani D, Chen J, Mateo-Babiano I, Bean R, Corcoran J (2020) Docked and dockless public bike-sharing schemes: Research, practice, and discourse. In: Curtis C (ed) Handbook of sustainable transport. Edward Elgar, London
Quah J (2003) Causes and consequences of corruption in southeast asia: a comparative analysis of Indonesia, the Philippines and Thailand. Asian J Public Adm 25(2):235–266
Redclift M (2005) Sustainable development (1987–2005): an oxymoron comes of age. Sustain Dev 13(4):212–227
Robinson J (2011) The spaces of circulating knowledge. In: McCann E, Ward K (eds) Mobile urbanism: cities and policymaking in the global age. University of Minnesota Press, Minneapolis, pp 15–40
Rose R (2004) Learning from comparative public policy: a practical guide. Routledge, New York and London

Sassen S (2001) The global city: New York, London, Tokyo. Princeton, NJ: Princeton University Press

Stead D (2012) Best practices and policy transfer in spatial planning. Plann Pract Res 27(1):103–116

Stead D, Banister D (2003) Transport policy scenario building. Transp Plann Technol 26(6):79–102

Temenos C, McCann E (2012) The local politics of policy mobility: learning, persuasion, and the production of a municipal sustainability fix. Environ Plann A 44:1389–1406

Thomas R, Pojani D, Leffernik S, Bertolini L, Stead D, van Krabbe E (2018) Is Transit Oriented Development (TOD) easily transferable to international contexts? Reg Stud 52(9):1201–1213

United Nations (2014) Handbook on Measurement, Reporting and Verification for developing country parties. Report, United Nations Climate Change Secretariat, Bonn, Germany. https://unfccc.int/files/national_reports/annex_i_natcom_/application/pdf/non-annex_i_mrv_handbook.pdf. Accessed 27 Oct 19

von Mises L (1944) Bureaucracy. Yale University Press, New Haven, Conn

Wolman H, Ford C, Hill E (1994) Evaluating the success of urban success stories. Urban Stud 31(6):835–850

Wolman H, Page E (2002) Policy transfer among local governments: an information-theory approach. Governance: Inter J Policy, Adm, Institutions 15(4):477–501

Wu I, Pojani D (2016) Obstacles to the creation of successful bus rapid transit systems: the case of Bangkok". Res Transp Econ 60:44–53

Chapter 4
Sustainable Urban Transport in Southeast Asia: Making It Happen

Abstract This chapter delineates the way forward. In the case study cities, it is universally accepted that there is a transport crisis. Knowledge about what sustainable transport interventions could be adopted locally also abounds. Implementation—rather than the transfer of ideas—is the problem. The chapter looks at the public policy and administration literature for suggestions on how to bring about implementation. Existing models suggest that, up to now, implementation of sustainable urban transport projects in Southeast Asia has been mostly 'symbolic' or 'experimental'. The goal for the future would be to move Jakarta, Manila, Kuala Lumpur, and Bangkok to 'administrative' and 'political' forms of implementation which carry less ambiguity of goals and means.

Keywords Urban transport crisis · Institutional barriers · Implementation

In Jakarta, Manila, Kuala Lumpur, and Bangkok it is universally accepted that there is a transport crisis. By now, it has become clear that private motorization is not the way to meet the travel needs of a growing population while also protecting the environment. Although local populations appear to be quite attached to the car (and the motorcycle), they are also exhausted and exasperated by transport problems such as congestion, pollution, unsafety, and general chaos, and increasingly demand better solutions.

Knowledge about what sustainable transport interventions could be adopted locally also abounds in all four capitals. Therefore, any recommendation on what needs to be implemented would be merely a cliché. The concept of 'sustainability'—contradictory as it is—has diffused to Southeast Asia, and has inserted itself in all policy documents. Between the Internet, study tours, technical assistance, and foreign degrees, all policy transfer actors—politicians, bureaucrats, consultants, technology producers and vendors—are well versed on the menu of sustainable transport options available internationally. 'Rich and famous' Western European and East Asian cities are the preferred role models here, with each city being renown for one or more aspects of their transport system (e.g., Amsterdam for cycling networks, Tokyo for rail transit, Guangzhou for BRT, etc.). Singapore, as the most advanced of ASEAN nations, is often an explicit or implicit benchmark (Table 4.1).

D. Pojani, *Planning for Sustainable Urban Transport in Southeast Asia*, The Urban Book Series, https://doi.org/10.1007/978-3-030-41975-2_4

Table 4.1 Summary of main findings

Drives	Drivers	Description	Direction	Deterrents	Degree
Diffusion of the sustainability concept	Individual political champions	Physical development	From the European 'Pentagon'	Financial/institutional/legal frameworks	Inspiration
Financial opportunities	Local experts	Implementation processes and incentives	From Japan	Corruption/clientelism/populism	Fragmented implementation
Urban boosterism and marketing	Planning student diaspora	Public participation processes	From Asian Tiger economies	Apathy/inertia	Transfer failure
Study tours	International technology mongers			Communication	
The Internet				Auto symbolism	
				Policy fragmentation	
				Local cultures	
				Local climate	
				Physical setups	

The transfer of *ideas* is not the problem. In fact, it has proceeded quite successfully (with the caveat that adjustment to the local context is needed in all cases). The problem is with putting the acquired knowledge into practice and making urban transport sustainability happen locally. Despite a substantial amount of help and funding from abroad (in the form of both loans and grants), the implementation of public and active transport projects (e.g., cycling networks, BRT, MRT, etc.) has been slow, patchy, or even non-existent. Many plans, the execution of which would have been highly desirable, have been prepared and then shelved for years. Policy measures that should complement the construction of physical infrastructure (such as congestion charging, value capture, minivan fleet renewal, etc.) have rarely been adopted too. In no case have integrated packages of sustainable transport solutions been implemented.

Southeast Asian capitals, like most cities throughout the world, are in a political and institutional logjam with regard to sustainable transportation planning and implementation. This is not a policy transfer issue. The implementation barriers would have been the same had the policy concepts been born domestically rather than having foreign origins (although international policies carry a higher status). As Southeast Asian countries have overcome the transition stage from poverty to middle income,[1] the nature of technical assistance (or supply-driven policy transfer) to the region should change. The focus should shift from visioning and planning to implementation. So the focus should be on the domestic urban planning sector (both transport and land-use) and on demand-driven policy transfer.

For ideas on how to unlock the situation, a look at the public policy and administration literature is useful. Matland (1995) has proposed a comprehensive implementation model, which is illustrated in Fig. 4.1. This model takes into account both political conflicts and policy ambiguity (of goals and means). When applied to the context of the Southeast Asian capitals discussed here, the model suggests that, up to now, implementation has been mostly 'symbolic' or 'experimental'. In other words, it has depended largely on the stream of actors that have been active and involved in the micro-implementing environment. Ambiguity, turf wars, corruption, and inertia have led to a proliferation of interpretations. The processes have been fluid and the results are unpredictable. Outcomes have varied a great deal in terms of quality, depending on the strength of the political and bureaucratic coalitions which have been in control of the resources.

The goal for the future would be to move Jakarta, Manila, Kuala Lumpur, and Bangkok to 'administrative' and 'political' forms of implementation which carry less ambiguity. While 'administrative implementation' in the field of sustainable transport tends to be the province of places with mature planning systems and a long history of rational planning, 'political implementation' appears to be realistic and in line with the top-down governance style and the existing models of decision-making in Southeast Asia.

[1] With Malaysia poised to reach 'developed' status soon.

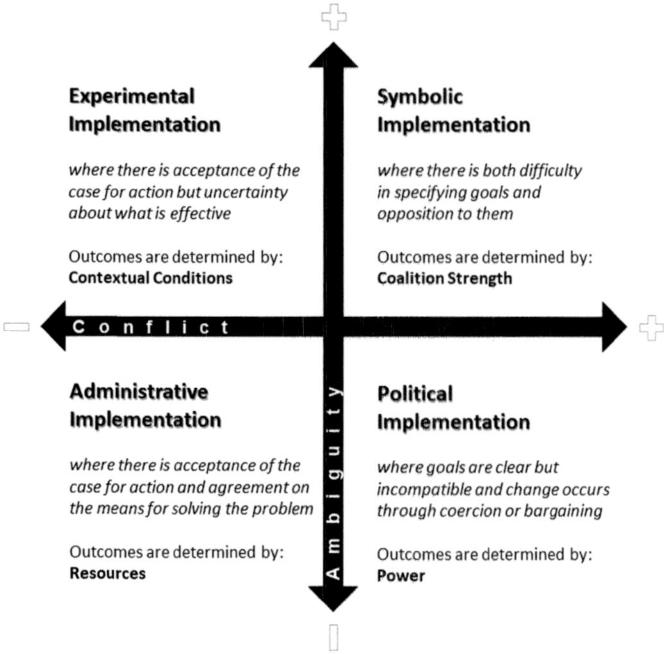

Fig. 4.1 The impact of political conflict and policy ambiguity upon implementation. Adapted from Matland (1995) and Hill (2016)

In addition to elected officials and career bureaucrats, most countries have constituent monarchs or traditional rulers. It is important to enlist their support as well—even if only symbolic. Southeast Asian monarchs, even where they have little governing power, often enjoy at least nominal cultural authority over the local populations. Also, their succession is often hereditary and thus they are protected from electoral turnovers. Successful implementation will also depend on the cooperation of urban transport actors, external to government, such as relevant constituency groups and civil society organizations who control vital resources. Sustainability champions will need to ensure that the process is not thwarted by opponents of particular policies. Meanwhile, governments need to be prepared to bargain agreements on means, and even employ 'carrots' and/or 'sticks', such as sanctions or grants. These are more effective where stronger monitoring and oversight procedures are in place (Matland 1995). While conflicts may occur, and compliance may not be automatically forthcoming, the key is to induce and support governments in their effort to exert their inherent authority in favour of urban transport sustainability.

References

Hill M (2016) Implementation: an overview. The Public Policy Process. Routledge, London, pp 224–226

Matland R (1995) Synthesizing the implementation literature: the ambiguity-conflict model of policy implementation. J Public Adm Res Theor 5(2):145–174

GPSR Compliance

The European Union's (EU) General Product Safety Regulation (GPSR) is a set of rules that requires consumer products to be safe and our obligations to ensure this.

If you have any concerns about our products, you can contact us on ProductSafety@springernature.com

In case Publisher is established outside the EU, the EU authorized representative is:

Springer Nature Customer Service Center GmbH
Europaplatz 3
69115 Heidelberg, Germany

Batch number: 10756378

Printed by Printforce, the Netherlands